The Sun and the Other Stars of
Dante Alighieri

A Cosmographic Journey through the Divina Commedia

Other Related Titles from World Scientific

The Enchantment of Urania: 25 Centuries of Exploration of the Sky
by Massimo Capaccioli
ISBN: 978-981-124-777-4
ISBN: 978-981-124-927-3 (pbk)

Our Celestial Clockwork: From Ancient Origins to Modern Astronomy of the Solar System
by Richard Kerner
ISBN: 978-981-121-459-2
ISBN: 978-981-121-531-5 (pbk)

Small Bodies of the Solar System: A Guided Tour for Non-Scientists
by Hans Rickman
ISBN: 978-1-80061-051-4
ISBN: 978-1-80061-060-6 (pbk)

A Cabinet of Curiosities: The Myth, Magic and Measure of Meteorites
by Martin Beech
ISBN: 978-981-122-491-1

The Sun and the Other Stars of
Dante Alighieri

A Cosmographic Journey through the Divina Commedia

Sperello di Serego Alighieri
Astrophysicist and descendant of Dante Alighieri

Massimo Capaccioli
Università Federico II, Naples, Italy

World Scientific

NEW JERSEY · LONDON · SINGAPORE · BEIJING · SHANGHAI · HONG KONG · TAIPEI · CHENNAI · TOKYO

Published by

World Scientific Publishing Co. Pte. Ltd.

5 Toh Tuck Link, Singapore 596224

USA office: 27 Warren Street, Suite 401-402, Hackensack, NJ 07601

UK office: 57 Shelton Street, Covent Garden, London WC2H 9HE

Library of Congress Cataloging-in-Publication Data
Names: Di Serego Alighieri, Sperello, 1952–   author. | Capaccioli, M., author.
Title: The sun and the other stars of Dante Alighieri : a cosmographic journey through the
    Divina commedia / Sperello di Serego Alighieri, INAF, Osservatorio Astrofisico di Arcetri, Italy,
    Massimo Capaccioli, Università Federico Secondo, Italy.
Description: Hackensack, NJ : World Scientific Publishing Co., [2022] |
    Includes bibliographical references and index.
Identifiers: LCCN 2021033764 | ISBN 9789811245497 (hardcover) |
    ISBN 9789811246227 (paperback)
Subjects: LCSH: Dante Alighieri, 1265-1321. Divina commedia. |
    Dante Alighieri, 1265-1321--Knowledge and learning. | Astronomy, Medieval. |
    Astronomy, Medieval, in literature. | Cosmography in literature.
Classification: LCC PQ4401 .D5 2022 | DDC 851/.1--dc23
LC record available at https://lccn.loc.gov/2021033764

British Library Cataloguing-in-Publication Data
A catalogue record for this book is available from the British Library.

For any available supplementary material, please visit
https://www.worldscientific.com/worldscibooks/10.1142/12507#t=suppl

Desk Editor: Ng Kah Fee

Typeset by Stallion Press
Email: enquiries@stallionpress.com

Printed in Singapore

Contents

Chapter 1

Introduction

The sciences, astronomy in particular, were of great importance for Dante. In the *Convivio* he writes: *"la scienza è ultima perfezione de la nostra anima, ne la quale sta la nostra ultima felicitade, tutti naturalmente al suo desiderio semo subietti"* (*Convivio* **I**, I, 1) [1] "science is the ultimate perfection of our soul, in which lies our ultimate happiness and which we all naturally desire". A little further he specifies: *"E questa (l'astronomia) più che alcuna de le sopra dette (scienze) è nobile e alta per nobile e alto subietto, ch'è de lo movimento del cielo; e alta e nobile per la sua certezza, la quale è sanza ogni difetto, sì come quella che da perfettissimo e regolatissimo principio viene. E se difetto in lei si crede per alcuno, non è da la sua parte, ma, sì come dice Tolomeo, è per la negligenza nostra, e a quella si dee imputare."* (*Convivio* **II**, XIII, 30), "And astronomy, more than any of the other sciences, is noble: it is prestigious and highly prized for its certainty regarding the celestial system, which is flawless since it is based on the most perfect and regulated system. And if anyone thinks that there is a weakness, then as Ptolemy says, that should not be attributed to science but to our neglect."

Therefore science is perfection and a source of happiness (Leonardo da Vinci will later say: *"Li omini boni desiderano sapere"* "Good men want to know"), and astronomy is the most noble and highest science. In the Middle Ages the culture of the few scholars ranged from the arts, such as poetry, to philosophy, from the natural sciences to theology. With his prodigious memory and his unique ability of interlacing different disciplines,

---

[1] Citations from Dante are given in Italian, followed by a translation in English. For the *Divina Commedia* we followed the English translation by R. M. Durling (1996).

Dante is one of the main representatives of this culture, as clearly proven by his many works, particularly by the main one, the *Commedia*.

The *Divina Commedia* — as it was called in a sixteenth-century edition which applied to the entire poem the adjective *divine* that had been given just to the *Cantica* of *Paradiso* by Giovanni Boccaccio in his *Trattatello in laude di Dante* (Small Treatise in praise of Dante) — is the story of a journey across the Universe, as it was understood at the time. Without doubt it is the most famous account of the three otherworldly realms, a work of science fiction *ante litteram* and also a cosmology and cosmogony treatise, written in poetic form.

The journey started with a descent to the center of the Earth — thought to be also the center of the Universe and the origin of gravity — and ended with the ascent to the Empyrean, beyond the stars, at the edge of the Universe. To tell us about his otherworldly experience, Dante made continuous reference to the culture of his time in the geographical, astronomical, philosophical, and theological fields. However, it must be said that in the Middle Ages the fragmentation of knowledge that in modern times has led to the rapid development of specialisations had not yet been implemented. For example, in Dante's time a familiarity with the sky was common even on a popular level, also because only the light of the celestial bodies illuminated the way for travelers at night. Dante's frequent astronomical references could therefore be grasped by readers of his time much more easily than by those of today; for this reason, in our book we offer a detailed analysis of these references, bringing them back to the framework of cosmological knowledge of the time.

Before we start describing and discussing the many astronomical references in Dante's works, particularly in the *Divina Commedia*, we provide a historical introduction, which will help the reader in putting the life and the works of the poet in the right context. We will then continue with a concise biography of Dante, focusing on the aspects that are relevant for the understanding of his works and, in particular, on those with cosmographic significance.

In order to better understand Dante's stances in geography and astronomy, it seemed appropriate to place them in relation to the earlier and contemporary relevant theories to which the poet consciously referred, at the same time underlining the differences with respect to those of today. Therefore Chapter 4 will deal with the developments of astronomy up to the time of Dante.

In the main part of this book, we will first discuss the most important temporal references encountered in the *Commedia* that allow us to reconstruct, among other things, the date and duration of Dante's journey. In analysing the cosmographic aspects in Dante's *Commedia*, we have adopted a different criterion from that favored by most scholars, who in their works have followed the order in which the various themes appear in the poem. Rather, we have chosen to divide the topics by astronomical subject, ordering them according to an increasing distance from the Earth, starting from the geography of our planet, then moving on to atmospheric phenomena and to the various celestial spheres, and ending with the overall structure of the Universe.

Finally, to help the reader immediately identify the treatment of a given cosmographic topic in the *Divine Comedy*, we have added an index of the corresponding passages in order of their appearance in the poem.

Large part of this book has already appeared in another book in Italian published in 2021 by GEDI Gruppo Editoriale S.p.A. with the title *"Il Sole, la Luna e l'altre Stelle. Viaggio al centro dell'universo dantesco"*, within a series of 5 volumes on Dante Alighieri promoted by the newspaper "La Repubblica" and edited by Marcello Ciccuto and Domenico De Martino.

Chapter 2

Historical introduction

2.1 The start of a conflict for the *auctoritas* and the *imperium*

Dante Alighieri (1265–1321) was born in the heart of the so-called Late Middle Ages (Early Modern History), a period of history conventionally included between the turn of the millennium and 1492: the year in which Columbus discovered the Americas; Lorenzo the Magnificent, icon of the Italian Renaissance, died in Florence; and the Sultanate of Granada, the last Spanish bastion of Islam, fell by the hand of the Catholic monarchs Ferdinand II of Trastámara and Isabella I of Castile. Three events that, for different and complementary reasons, marked the start of the Modern Age.

A watershed between two epochs of great social, cultural, and scientific development of the Old Continent, the Middle Ages is made to begin in 476 BC with the fall of the last Western Roman Emperor and the coronation of the barbarian general Odoacer as a Roman patrician. Actually, the deposition of Flavius Romulus Augustus (also called Augustulus due to his young age and historical irrelevance) only marks the formal birth of a process that started at least two centuries ahead: the disintegration at a continental scale of the old order with the coming on the scene of new actors not homologated to the Roman standard. Men more vital than those whom a long history of power had already worn out. The armed penetration of the Slavic and Germanic populations into the Balkans, Western Europe, and Italy, in turn driven by the exuberance of the Huns, determined a first fragmentation of the territorial, political, linguistic, and cultural unity built and maintained for centuries by the Roman legions.

Already violated for the first time by Alaric's Visigoths in 410, the city of Rome was heading towards a disastrous decline marked by the loss of the ancient and glorious identity of *caput mundi*. The crisis was much deeper than a simple change of ownership of the imperial throne (previously experienced many times, starting from when the barbarian Maximin of Thrace had taken over the purple in 235). It contemplated the loss of authority of the senate, definitively dissolved in the VII century; the increasingly frequent exposure to military invasions and looting; the relaxing in the care of the public good; the depopulation of the countryside with the disappearance of vital food resources and the consequent demographic collapse,[1] worsened by repeated plagues and unfavorable climatic changes;[2] and, with the Byzantine[3] invasion of Italy under Justinian at the end of the VI century, the transfer, even symbolic, of the government of Italy to Ravenna, on the delta of the river Po, with the migration of the military and senatorial aristocracy to the capital of the newborn Exarchate.[4] In the ancient capital, the absence of the secular power gave a chance to the growth of a new dominant figure who legitimately encamped a divine investiture, the bishop of Rome, named pope since the III century AD: successor of the apostle Peter who had moved to the Eternal City to found there the Church of Christ.[5]

---

[1] The population of Rome, which in the Antonine period (II century) had reached the peak of about 1.5 million people, after the sack of the Vandals (year 455) reduced to 200,000 inhabitants. This number was further halved at the end of the Gothic War.

[2] Starting from the middle of the II century AD, the Mediterranean climate, which had remained warm and temperate for a long time, underwent a progressive cooling down until approximately the year 1000, accompanied by the appearance of new pathogenic organisms that ignited devastating pandemics in Rome and in the empire. Events such as the so-called Cyprian plague (249–262 AD) and the epidemic that exploded in the Justinian age (541–542), concomitant with the little ice age of late antiquity associated with the occurrence of very violent volcanic eruptions, have undermined the economic and demographic foundations of the empire, exposing it to the deadly blows of the barbarians.

[3] The term "Byzantine" used to represent the Eastern part of the Roman Empire, of predominantly Greek culture, separated from the Western and more exquisitely Latin part by Theodosius in 395 AD, was introduced only in the XVIII century by the Illuminists. Actually, the Byzantines called themselves Romans or Romei.

[4] *Exarchatus Ravennatis* is the name given by the Byzantines to the governorship of the Italian territories placed under the control of Constantinople upon the reconquest of the peninsula in the VI century, divided into ducats ruled by a duke (*dux*): the Exarchate centered on the capital Ravenna, Istria, Veneto, Liguria, Pentapolis (cities of Rimini, Pesaro, Fano, Senigallia, Ancona), Perugia, Rome, and Naples.

[5] Actually, all the early medieval cities experienced the absence of civil power. Usually it was the figure of the bishop to make up for the structural deficiencies. He administered justice, organized militias, appointed officials, and in some cases crowned sovereigns. Basically he ruled. This also happened in Rome, of course. But here the local bishop

Between 535 and 553, the Italian peninsula was shattered by the war between the Goth invaders[6] and the Byzantines, determined to (re)conquer the western half of the Roman Empire. In a condition of total anarchy, the popes, while not yet formally invested with power over the city (in this period, they were still simple patriarchs of the imperial Church), began to increase the properties of the Church through donations of land and money from followers and pilgrims drawn to Rome by the tomb of Peter. Greater wealth meant greater power. The pontiffs acquired their own militia, appointed magistrates to exercise justice, and managed the city, providing what was needed for the material sustenance of an impoverished and riotous mob. Supported by a new class of landowners by then urbanized, they were expressed by the leaders of powerful families and by representatives of the people, with a format that only superficially recalled the practice for electing consuls in vogue in ancient Rome. The popular consent was manipulated by a litigious oligarchy in the making.

The complete victory of the imperial army over the Goths had short-lived effects. Internal brawls in the Byzantine court, together with the pressure exerted by the Avars to the east, pushed a Germanic population that had settled in Pannonia to invade Italy. Led by their king Alboin, in 569 the Lombards easily conquered the northern regions of Italy without encountering obstacles from the imperial forces, and fixed in Pavia their capital. They also reached the central-southern parts of the peninsula where they founded the duchies of Spoleto and Benevento. Alboin was crowned *dominus Italiae*, king of Italy. Feeling himself surrounded by the barbarians, the pope asked Constantinople for help that the *basileus*, engaged militarily on his eastern front, was unable to provide. So the pontiff addressed his request to the Franks, another Germanic people who had by then established a kingdom in present-day France. This move had no practical consequences but it was almost a dress rehearsal of what would become the *modus operandi* of the papal diplomacy in following centuries.

Meanwhile, in year 602 the usurper Phocas had ascended the throne of Constantinople. For his intrinsic weakness but also because he was in tune with the social policy inaugurated by pope Gregory the Great, a small and

---

felt himself the heir to a universal power directly given to Peter by Christ with the words "*Feed my sheep*" (John 21, 15–17).

[6] The Goths, federates of the Byzantine Empire, had been sent to Italy by the emperor Zeno to drive out Odoacer. Conducted by Theodoric I, they arrived in 488, settled down, and created a kingdom that was soon disliked in Constantinople. Thus, in 535 Justinian decided to reconquer the Italian peninsula.

sickly man endowed with great moral strength and unshakable faith, Phocas proclaimed the Roman pontiff "head of all churches", while retaining for himself the right to approve his appointment. This was a very important recognition, which legitimized the growing claims of protagonism of the bishop of Rome and paved the way for creeping conflicts of power disguised as theological questions. It soon became clear the urgent need to exorcise the possibility that the pontiff would take over the imperial power to the point of becoming its dispenser. A first hint of how real this danger was came in the year 619. In his attempt to seize the title of Western Emperor, the rebel exarch Eleutherius tried in vain to be crowned by Boniface V, thus recognizing to the Roman pontiff, albeit for reasons of necessity, a role that could be read as a sign of a true supremacy of the church over the Eastern Empire. But at this point the *basileus* was about to pass his hand to a new player in the confrontation with the pope.

2.2   The era of the Carolingian kings

In 751 the imperial capital Ravenna fell to the hands of the Lombards, who threatened to point the finger at Rome. As usual, pope Zachary invoked the help of the Byzantine emperor. Powerless to provide any support, the latter let the pope turn to the Franks' king Pepin the Short. It is important to underline that this freedom of negotiation represented a further danger-ous handing over of the imperial power. In 754 Pepin crossed the Alps. The move of the Frankish king opened on a period of clashes, armistices, agreements, political marriages, and betrayals, that ended tragically for the Lombards. Instead, with a first seed of the State of the Church (Exarchate and Pentapolis), this conflict marked the birth of the temporal power of the Roman pontiffs, certified by a false document.

According to a paper known as *Constitutum Constantini*, two years after the victorious battle of the Milvian Bridge in 312, the emperor Con-stantine the Great had conferred pope Sylvester I and his successors the civil jurisdiction over Rome, Italy, and the entire West: an act heralding tragic consequences, would have later sentenced Dante (*Inf.* XIX, 115–117) who, while feeling closer to the ideals of the poor church of the origins, yet believed in the authenticity of the document:[7]

---

[7]In the first half of the XV century, the Italian humanist Lorenzo Valla proved by a linguistic analysis that the document had been prepared in the VIII century by the Vatican chancellery.

> *Ahi, Costantin, di quanto mal fu matre,*
> *non la tua conversion, ma quella dote*
> *che da te prese il primo ricco patre!*

"Ah, Constantine, not your conversion, but that dowry which the first rich father took from you, has been the mother of so much evil!"

For his part, the Frankish king Pepin was proclaimed *patricius Romanorum*. The transition of the title due to the Byzantine exarch symbolically indicated a passing of the baton to a new power.

Meanwhile, the city of Rome was resurrecting, owing to the activity of great popes such as Adrian I (772–795). Heir of a powerful family of the military aristocracy and proudly aware of the great past and historical role of the city, Adrian took care of restoring to the *urbs* its ancient monumental aspect. He promoted the construction of new civil and religious buildings and the recovery of a better life quality by repairing, for example, aqueducts and sewers, and he improved safety by reinforcing the defences. With the rebirth of a city class, some families began to affirm themselves: for instance, the dukes of Spoleto and the Crescenzi. From the *insulae* where arrogantly soared the towers of their fortified residences they would have dominated, in perennial conflict with each other, the Roman political scene in the following centuries.

Skilled diplomat, Adrian was able to navigate between Franks and Byzantines, ensuring the protection of the former and, in crucial moments, the neutrality of the latter. However, he was also the pope who exalted, with his behavior, one of the great vices of the Lateran court: nepotism that in the centuries to come would have stained the throne of Peter with infamy, to the point of becoming, together with simony,[8] one of the reasons for a moral (as well as political) revolution.

Adrian I had his sword in Pepin's son, Charlemagne. The new king of the Franks wiped out the Lombards from Italy, causing their definitive disappearance from the scene. He also assumed on himself the title of king of the Lombards (which immediately passed to one of his sons). This fact displaced the pope who, counting on a promise, intended to expand the possessions of the Church. Despite this turnaround, in the year 800 Charlemagne was crowned emperor by Adrian's successor, Leo III, during the Christmas

---

[8]Sale or acquisition of church offices and roles and of spiritual goods. The name derives from the request made by Simon Magus to the apostle Peter to acquire, in exchange for money, the thaumaturgical faculty of the Holy Spirit. Dante devotes canto XIX of the *Inferno* to the simoniac popes.

mass in St. Peter's: a formality that marked a turning point in the history of Europe. More than three centuries since the fall of Romulus Augustulus, in fact, and just when the weak empress Irene of Athens was sitting on the throne of the other half of the ancient Empire, the title that belonged to the Caesars reappeared in Rome, credited to a powerful sovereign by the pope. Ideally, the *urbs* took back its role of *caput mundi*.[9] Other forgotten titles and institutions also reappeared, such as consul, flanking the Byzantine duke and count (from Latin *comes*), and senate, to indicate then the assembly of the families of the new patriciate, those who claimed the right of electing the emperor (through the pope).

After extending his dominion over the Lombard lands of Northern Italy and over much of present-day Germany, Charlemagne seemed to succeed in his attempt to rebuild a universal power owing also to the synergy established with the pontiff. The experiment, however, did not survive him for long. In addition to the usual difficulties in the dynastic succession and the pressure exerted by Normans and Slavs at the borders, his project was hampered by his own idea of a new political, economic, and social order based on granting life-long benefits (fiefdoms) to vassals in exchange for military loyalty. The Carolingian vassal-beneficiary system had been conceived to strengthen the central power by establishing personal and direct connections between the king and the aristocrats. It contributed instead to crack it by creating centrifugal autonomist ambitions with dynastic purposes and competing economic and military forces. Settled in their castles (from the Latin *castrum*, fortified camp), the local lords, lay or ecclesiastical, defended their possessions (no longer perceived as royal concessions, but as private properties) with small personal armies: a mosaic of resources and ambitions controlled with increasing difficulty by the king or the emperor. The

---

[9] A leading role that, for Dante Alighieri, is due to the divine will. Realized by Aeneas, chosen to found the noble Rome and its empire, this design tended to establish the holy see for the successors of the first pope (Peter):

> ch'e' fu de l'alma Roma e di suo impero
> ne l'empireo ciel per padre eletto:
> la quale e 'l quale, a voler dir lo vero,
> fu stabilita per lo loco santo
> u' siede il successor del maggior Piero.
> (*Inf.* II, 20–24).

"for he in the Empyrean heaven had been chosen to be father of mother Rome and her empire: and Rome and her empire, to tell the truth, were established to be the holy place where the successor of great Peter is enthroned."

only other force capable of a centralized order was the Church. Although deprived of its legions, Rome relaunched through its bishop its political influence on the Western world, in continuity with the Latin tradition.

2.3 Otto I of Saxony and the birth of the Holy Roman Empire

The legacy of the Carolingian Empire was collected in the following X century by the German prince Otto I of Saxony, known as the Great. Valent leader, skilled politician, fine strategist and organizer, patron and protector of the arts and letters, he was able to defend the eastern and southern borders of his lands by stemming the advances of the Hungarians, Magyars, and Slavs. After transforming Germany into a continental power, hegemon over the Western Christian world, he launched an expansionist campaign in Italy. While at the end he did not attain the expected success, already in 962 he was crowned emperor in St. Peter's by pope John XII in exchange for the commitment to defend the Apostolic See. The revival of the Holy[10] Roman Empire under the aegis of the Roman Church had an important consequence. For the next five centuries or so, the peninsula would have to deal with the presence and interference of the Germanic kings in the government and in matters of faith, with a long period of fights, even fratricidal, between the partisans of the two new poles of power, the German emperor and the Roman pope.

By proposing himself as the ideal successor of the Caesars, almost in continuity with Alexander the Great, Otto had momentarily taken over the pontiff, limiting his temporal sovereignty to the Roman territory only and subordinating Italy, *"'l giardin de lo 'mperio"* (the garden of the empire) says Dante (*Pur.* VI, 105), to the Germanic crown. But this was only the first act of a long and bloody saga; those immediately following contained *in nuce* the elements characterizing the history of Italy and consequently, albeit three centuries later, the misadventures of Dante Alighieri. Let us outline them, limiting ourselves to the relations between the empire and the papacy.

While Otto was engaged in the north of the peninsula to contain the claims to the throne of Italy, he was informed that the pope was plotting with his enemies against him: prelude to the establishment of the

---

[10] Adjective added two centuries later by Frederick I Hohenstaufen to mark the merging of two powers into one.

practice of *divide et impera* (divide and command) adopted over the centuries by the bishops of Rome (to compensate the lack of military forces of papacy). Furious at the betrayal, the emperor rushed to Rome in arms. It was his second *Romzug* or *expeditio italica*, as these descents to Italy of the German emperors were then called, to designate a sort of usual practice. Pope John XII, who was the son of a corrupt nobleman holding power in the Eternal City and in turn a champion of debauchery — "*the worst possible of everything*", cardinal Robert Bellarmino would have labeled him six centuries later —, had time to flee and was put under indictment *in absentia* with very heavy charges: perjury, murder, incest, and blasphemy. Actually, Otto had already criticized the pope in unsuspecting times, when he was still seeking his collaboration. Called to justify himself, John XII, though very young, reacted firmly. He excommunicated the emperor and all the members of the synod who had pointed at him the finger. Otto responded with force to the act of force.[11] He deposed John and had a new pontiff appointed to his liking. The choice fell on a layman: the sober, pious, and docile head of the Lateran Chancellery, who took on the name of Leo VIII. This heavy interference in ecclesiastic questions was justified by an act that the emperor had issued in 962, the *Privilegium Othonis*, stating his right to choose and confirm the pope. But the faction of the deposed pope, duly instigated, rose up against the imperials with such a violence that Otto was forced to flee, protected by his heavy cavalry that saved him by charging the crowd. A few days later, even the innocuous Leo VIII had to leave Rome speedily while a triumphant John XII returned to the Lateran. What followed offers once more an exemplary insight into the history of Italy in the centuries to come.

In 964 John XII died. Disregarding the Ottonian privilege, the Romans elected in total autonomy a new pope, Benedict V. Although he was an irreproachable and pious man of letters, yet his election was a matter of insubordination. The emperor had again to descend to Italy in arms to reassert his authority. Rome was besieged and conquered by starvation, the rioters severely punished, and Leo VIII was restored by force to the throne of Peter. But after only one year he too passed away. Made cautious by the harsh lesson received, the Romans asked Otto, who in the meantime had returned to Saxony, to allow them to call back into service the deposed pope

---

[11]With the exclusion from the sacraments, the excommunication of a sovereign entailed the loss of the divine right from which his royal power was made descend, freeing the subjects from the obligation to obey.

Benedict. The emperor refused, making it a matter of principle, without however deciding to assert his *Privilegium* for fear of other revolts.

The unexpected death of Benedict temporarily cleared the impasse. Otto dropped his choice on a person loyal to him, but the new pope, John XIII, soon became unpopular because of his authoritarian behavior towards the Romans and, within a year, he too was forced to leave the city quickly under the pressure of an angry crowd. It was really too much for the patience of the German sovereign, who came down to Rome for the fourth time to restore imperial authority with iron and fire, fiercely assisted in the cleaning operation by John XIII himself. On the occasion, he also wanted to ensure the succession of his fourteen-year-old son in second bed, Otto II, associating him with the throne and making him in turn nominate emperor. He also found him a wife, the Byzantine princess Theophanu, chosen with a specific political purpose: to acquire peacefully some rights on the territories occupied by Constantinople in Southern Italy (Apulia and Calabria) and thus unify the peninsula under a single crown, with the connivance of the surviving Lombard dukes. This program was in total contrast to the papal designs.

Returning to Saxony, in 973 this great king suddenly died of poisoning from spoiled food, leaving triggered the trap on which even Dante would have stumbled well.

2.4   The first Saxon dynasty and Italy

In 980 Otto II, son and heir of Otto the Great, descended in Italy. His declared aim was to put an end to the incursions on the continent by the Arabs of Sicily. He could count on the support of the Lombard dukes of Benevento and on the benevolent neutrality of Constantinople because of his marriage with the imperial princess Theophanu. But the hidden purpose of his military enterprise was another one. Otto II planned to extend the imperial control over the whole peninsula, including the reconquer of Sicily itself, justifying it as a return of the island to the bed of Christianity: an argument that bound the pope to accept *obtorto collo* the deprecated hypothesis of a complete encirclement of Rome by the German emperor.

The expedition, however, did not meet the hoped-for success. The German army was defeated by the Arabs in a bloody battle and Otto was forced to retreat to Rome where he died at just twenty-eight, perhaps of malaria, comforted in his last hours by the pope himself. This was the tragic end of his lone *Romzug*, which however should not make us think of a disinterest of the emperor in the affairs of the Eternal City. Through his delegates, Otto II in fact had to intervene indirectly in the dramatic story of the antipope Boniface VII.

Former cardinal deacon of John XIII, Boniface had seized the throne of Peter by deposing the legitimate pontiff Benedict VI (hence the appellative of antipope, as he was appointed pope with a pontiff still in office), then having him strangled in prison in order to prevent any attempt at restoration. The murder had however sparked popular indignation and Boniface was forced to leave Rome hastily and take refuge in the Byzantine dominions in Southern Italy. He then asked for hospitality in Constantinople where the *basileus*, who badly digested the expansionist policy of his German "colleague", welcomed him willingly. In 980 Boniface returned to office. Taking advantage of the absence of the emperor, who had to leave Rome to quell some internal turmoil in Germany, he was able to take over again the Lateran with the help of the Crescenzi, a powerful baronial family of Byzantine supporters who controlled the Roman plebs. The legitimate pope Benedict VII, who in the meantime had been elected with imperial approval and was therefore unwelcome to the Romans, was captured and assassinated. It was worse than a crime: it was a mistake. The same people who had risen against Benedict, then wanted to avenge his death. Thus Boniface was again forced to escape to Constantinople and a Lombard bishop, favorite of Otto II and in particular of his wife Theophanu, was raised to the throne of Peter with the name of John XIV.

But in 983 the new pontiff found himself helpless against the renewed assault of the irreducible Boniface, supported by the *basileus* and by the armed bands of the Crescenzi. The emperor, who was protecting him, had suddenly died and his widow had to move to Germany together with her three-year-old son, the future Otto III, committed to facing the dynastic claims of an ambitious relative, Henry II of Bavaria.[12] Captured and imprisoned, pope John XIV was suppressed and his body hung in the stands of the Castle of the Holy Angel. This was however a sort of Pyrrhic victory for the antipope. Boniface VII was able to reign one year only, possibly assassinated by a palace conspiracy.[13] He was succeeded by the cultured Roman presbyter John XV, elected with the favor of the German court.

---

[12]He was the son of a younger brother of Otto I and thus a cousin of Otto II.

[13]In the popular memory he was renamed *"Malefatius"*, to remark his bad reputation among the people.

The shameful joust of popes and antipopes was repeated during the reign of Otto III. The very young orphan had been elected king of the Germans immediately after the death of his father, in 983, and placed under the tutelage of his mother, the empress Theophanu, once she had managed to frustrate the attempt to seize regency by Henry the Quarrelsome, cousin and rival of Otto II. Educated by the most learned prelates of his time, Otto grew up in the myth of Rome and Byzantium and of the universal role of the empire, and in the contempt of the barbaric world of the Germanic tribes from which his fathers descended. His first intervention in Italy was provoked by the revolt of the Romans who expelled John XV, no longer protected either by the empress, who had died in the meantime, and by the Crescenzi. The pope took refuge in Roman Tuscia (Tuscany) under the protection of the margrave Hugh[14] who in turn invoked that of the young German sovereign. It was a triumphal *Romzug*. In 996 Otto was crowned king of Italy in Pavia and emperor in Monza. He was then received with all honors in Rome. There, on the death of John XV, he had his cousin Bruno of Carinthia, first German pope, elected to the throne of Peter with the name of Gregory V. It was a potentially winning move. The two cousins had divided all the power between them: Otto had reserved the temporal power to himself, reaffirmed by a further coronation as emperor at the hands of his cousin-pope; and Gregory V assumed the power over souls and the role of representative of the empire in Rome.

Persuaded to have most troublemakers among the Roman nobles under control, Otto returned to Germany trusting in the effectiveness of the German troops left to protect the pontiff. But once again, despite the warnings received for previous betrayals, the urban mob forced the German pope to flee. In his stead, the antipope John XVI was appointed by the Crescenzi with the favor of the Byzantine court. Otto, committed to defending his Balkan borders, had to wait one year before reacting firmly. On his descent to Rome, he stormed the Castle of the Holy Angel where the leader of the Roman rebels was perched and, having captured the ambitious antipope, had him tortured and killed. Gregory returned to the Lateran where he reigned until his death, which occurred in 999 not for natural causes. He was just 27 years old. The cousin, who was still in Italy, returned to Rome

---

[14] He was a relative of Otto, who ruled Tuscany, Spoleto, and Camerino. His figure has some relevance for our story in that, in the third quarter of the X century, Hugh moved his marquis residence from Lucca to Florence, thus proving the growing of importance of the city on the Arno river.

and imposed as the new pontiff his learned tutor, Gerbert of Aurillac, the first pope of French nationality who took the name of Sylvester II. It was a symbolic and substantial choice[15] towards both a reform of the papacy and a restoration of the ancient empire, marked by the adoption of Greek and Latin as official languages and by the move to Rome of the capital. It lasted the space of a morning, though. In 1002, at the age of twenty-two, Otto died, perhaps he too of malaria contracted in Italy, while he was about to forge a renewed alliance with the Constantinople through his marriage to a daughter of the *basileus*.

Lacking a direct successor and in the absence of rules to manage such a situation, Henry II, duke of Bavaria, came forward. His father had unsuccessfully tried to contend for power with Otto II and then to replace the queen mother Theophanu as regent in the name of Otto III. A pious as well as shrewd and determined man, Henry consolidated his authority in Germany by revolutionizing the management of the territory, no longer entrusted to high feudalism, always ready for revolt, but to lesser-ranking counts and bishops of his own nomination. Although little interested in Italian affairs, he too had to go down to the peninsula three times. His first *Romzug* was to regain the title of king of Italy with a successful military campaign, which was followed by a second descent to settle the usual dispute between pretenders to the throne of Peter.

In Rome, power was gradually passing from the Crescenzi family to that of the counts of Tusculum, who in 1012 managed to express their own pope, the energetic Benedict VIII. The Crescenzi, who demanded a sort of monopoly on the appointment of popes, reacted immediately nominating yet another antipope, Gregory VI. This time it was the pope in office who prevailed. He chased away the rival who however, not giving up, even crossed the Alps to invoke Henry's help. Pragmatically, the German chose Benedict because he recognized his reforming spirit and the ability to hold the turbulent city of Rome with an iron fist. Grateful, in 1014 Benedict VIII crowned Henry emperor of the Romans in St. Peter, reconfirming him all the rights listed by *Privilegium Othonis*. For a couple of years and until Benedict's death, the two governed the Western world in perfect agreement, sensitive to the winds of reform of the Church that were blowing strongly from the Benedictine Abbey of Cluny in Bourgogne. Henry died in 1024,

---

[15]The first pope naming himself Sylvester was also the one who had baptized Constantine the Great and who, according to tradition, had received the famous "donation".

after a third descent in Italy to put down the endemic revolts in the territories of the ancient Lombard kingdom, which though, upon hearing of his death, again tried to violently raise their heads. The German dukes, instead, remained calm, proving that the last king of the Saxons had been able to set up a satisfactory management of power.

2.5 The Salic emperors and the Investiture Controversy

Once again a Saxon ruler had disappeared without direct heirs. Starting from a second-rate position in the elite of the high Germanic aristocracy, Conrad of Franconia managed to rise to power and reign over a vast territory. The *pax romana* set by Henry II remained with him too. Conrad II was king of the Franks from 1024, king of Italy from 1026, and emperor of the Holy Roman Empire from 1027 until his death in 1039. He was the first exponent of a new dynasty called Salian from the name of a tribe of the Franks which in Roman times had settled in present-day Holland. Unlike Henry the Saint (made a saint immediately, together with his wife Cunigunde, who had held the reins of the state in the interregnum period following the death of her husband), Conrad had a secular and highly pragmatic mentality. In order to reinforce the central power, he weakened that of the great feudal lords to the advantage of the largest and most faithful audience of the vavasours. Like his predecessors, he was very attentive to reaffirm the imperial authority in Italy and to try to keep at bay the independence ambitions of some Lombard cities; in particular Pavia, former capital of the Lombard kingdom, which in the renewed period of trade came to be in a fortunate geographical position.

Between 1027 and 1028, while Conrad was staying in Rome to receive the imperial crown from the pope, an important event took place, fraught with consequences for the history of Italy and for Dante Alighieri's life. Ernest of Swabia,[16] first bed son of the wife of Conrad and of the late duke of Swabia, rebelled against his stepfather. Allied with a Swabian nobleman, Guelph of Welfen,[17] the young man invaded Burgundy while his ally put the ancient city of Augusta to fire and sword. Conrad hastily returned to his homeland and defeated them both. The stepson was imprisoned and Guelph deprived of his possessions, forced to return the spoils of the sacking of Augusta and exiled. This settling of scores was not different from

---

[16]Cultural, historic, and linguistic region in southwestern Germany.
[17]In arcaic German the word "Welf" means young wolf.

thousands of other similar episodes in those difficult times, but it gave rise to a long lasting conflict between Guelphs (Welfen) and Ghibellines (Hohestaufen; see later) destined to inflame the warm lands of Southern Europe for centuries.

A second episode deserves attention for its consequences. Aribert, a powerful archbishop of Milan and a faithful ally of the emperor — he had been the one crowning Conrad king of Italy —, had come into conflict with the small vassals of his vast fiefdoms, who demanded more freedom and better protection. In 1036 the growing discontent resulted in rebellion, which Aribert tried to extinguish by force. Finding himself in difficulty, the bellicose prelate invoked the help of Conrad who thus went down to Italy, settling in Pavia. But things turned out differently from what Aribert had imagined. Showing all his cynical realism, instead of taking the side of his faithful ally, the emperor used the opportunity to get rid of a bulky and potentially dangerous figure and had the man of God arrested; but the bishop managed to escape from prison.

This time the events turned in a different direction from the emperor's plans. Aribert's conviction was received by the people of Milan as an insult and had the effect of regrouping around the archbishop and the new symbol of municipal irredentism, the so-called *Carroccio*, the nobles of all ranks and the people, determined to oppose the arrogance of the foreign ruler. Conrad responded by besieging Milan. He also ordered pope Benedict IX to excommunicate Aribert, and even tried a *captatio benevolentiae* towards his vassals in the form of a *Constitutio de feudis* that made all fiefs hereditary and inalienable (not just lifelong possessions). But Milan did not fall or bend. Three years later, in 1039, Conrad died, struck down by gout.

He was succeeded by his son, Henry III known as the Black, former duke of Bavaria and Swabia and king of Burgundy, who became emperor of the Holy Roman Empire in 1046, when the freshly appointed pope Clement II placed the crown on his head. The German sovereign had come to Italy to resolve the intriguing tangle created by the election of three different popes by as many powerful Roman families. The story is this. In 1044, pope Benedict IX, expression of the Tusculani, was expelled from Rome at the instigation of the Crescenzi and in his place an antipope was appointed, with the name of Sylvester III. A few months later Benedict IX managed to reconquer the Lateran, but then ceded the title for a large sum to John Gratian, his godfather, who took the name of Gregory VI. It was really too much: the presence of three popes alive and the exhibition of the shameful market of the throne of Peter required a provision that would restore dignity to the Church. This was a request of the Cluniac monks whom Henry had approached thanks to his second wife Agnes of Poitou. But it was also an important matter to settle for the emperor himself, because the scandal discredited one of the two poles of that universal power on which his authority also rested. A council was called in Sutri, a town north of Rome chosen to avoid interference from the opposing parties of the city, which the sovereign himself wanted to preside. The three popes were declared lapsed (with different formulas, in formal compliance with canon law and in order to avoid yet another antipope), and a Saxon pope was elected: the former chancellor of the emperor and very loyal to him, who named himself Clement II.

But the story did not end there. After reconfirming the imperial right to intervene in the election of the pope, Clement died. Since Gregory was still alive, some voices were raised to ask that he were reinstated in the office of pope, but the emperor refused. Fortunately, Gregory too died shortly thereafter, freeing the emperor from an embarrassment. A Bavarian pope was thus elected, confirming the Germanic occupation of the papacy. Damasus II lived for a few months, almost always far from Rome because the irreducible Benedict IX had once again taken control of the city, from which he had to be forced out.

In 1056 Henry III died suddenly leaving as designated heir his six-year-old son, Henry IV, had by his wife Agnes of Poitou. During the long interregnum, until the coming of age of the new sovereign, the political situation in Germany degenerated. Agnes, who had proposed herself as regent, was forcibly ousted by the bishops of Cologne and Hamburg-Bremen, who also took care of the education of the child. Royal and imperial power gradually lost its strength. Finally, in 1065, the fourteen-year-old Henry IV was declared of age (decorated knight, it was then said). In possession of its full powers, the very young *rex Romanorum* initiated an action to contrast the autonomist tendencies of the imperial princes, trying to maintain a balance between all the forces in the field. But his authoritarian attitude, in a context of weakness within the Germanic nation, ended up spoiling his relations with the pope, just at the moment when a great pontiff, Gregory VII, sat on the throne of Peter.

The *casus belli* of that dispute, which would later be labelled the Investiture Controversy, another source of calamities in the centuries to come, was above all the heavy interference of Henry in the appointment of the bishop of the Milan diocese. A weak pope would certainly have bowed his head in the face of the arrogance of the emperor. Gregory, a severe reformer of the role, rules, and customs of the Church, was instead a warrior. He was fully persuaded of the supremacy of his office over any other earthly power,[18] and therefore excommunicated the emperor, declaring him lapsed. This was a bold counter-move to balance a similar provision by Henry (who had already had the German bishops declare the decadence of the pope for treason), taken in a moment of particular weakness of the emperor, committed to stifling the revolt of the Saxon dukes. The excommunication — it should be remembered — freed the subjects from the due obedience to the sovereign and therefore legitimized the action of the rebels.

Finding himself on the ropes, Henry IV had to surrender and to humble himself in order to obtain papal pardon. In a cold January of 1077, he and his wife Bertha of Savoy, barefoot as pilgrims and with ashes on their heads, for three days waited a sign from the pope at the door of the castle of Canossa, in upper Tuscany, where the Gregory VII was a guest of the powerful marquise Matilda of Tuscany. She was a noble, powerful, and spirited *Markgraf* of Lombard origins and absolute fidelity to the Church, whose fiefdom insisted on most of the north of Italy; a character whose subsequent actions would have changed the history of Italy. The "Walk to Canossa" marked the apex of the rise of papal authority, plastically represented by the humiliation of an emperor begging the pontiff to restore his lost authority. In any case, Gregory cancelled the excommunication but not the deposition of Henry, who thus remained *"among those who are suspended"*, to use the words of Virgil in Dante's *Commedia* (*Inf.* II, 52).

The tug-of-war between the empire and the Church was revived in 1080. Henry IV, who after the pardon had taken back with great effort the throne threatened by the duke of Bavaria, was again declared fallen by Gregory. This time it was the emperor who responded with an aggressive counter-move: he had the German bishops faithful to him declare Gregory's decadence and, to make the provision effective, he appointed a man of his as (anti)pope, who took the name of Clement III. This time the balance hung in favor of the German sovereign. His army managed to defeat the troops led by Matilda of Tuscany that the pope had sent to drive out the antipope Clement from Ravenna where he had settled. Henry then believed that the

---

[18] Gregory VII was the author of the *Dictatus Papae*, a collection of 27 propositions that sanctioned the rights of the Roman pontiff, including that of conferring power on the emperor and eventually revoking it: an authentic reversal of the theses of *Privilegium Othonis*. Obedience to the Church had to be absolute, under penalty of heresy and excommunication.

time had come for him to wear the crown of Italy and went to Pavia, but he was caught off guard by a further revolt of the German princes who elected a new king. Instead of hastily returning to Germany, the emperor decided to march on to Rome first. He intended to solve the pending matter with the rocky pontiff once and for all; but, failing to win at the first assaults, he left a delegate to continue the siege of the Holy City and took the way home.

He returned the following year and the next one, entering the city without ever being able to storm the Castle of the Holy Angel where the pope had perched. He then had an antipope appointed, who immediately crowned him emperor. That done, he turned back home to face and subdue the rebellious dukes of Swabia and Saxony. The pope, instead, was freed by the soldiers of Robert Guiscard,[19] duke of Apulia and Calabria: a skilled adventurer who from the first half of the XII century had settled with his Normans in Southern Italy.

2.6   The Normans of Italy

At the beginning of the X century the "men of the North", peasants and fearless navigators, had left the cold lands of Scandinavia and Denmark to settle in present-day Normandy, from where they took off to successfully invade various regions of continental and insular Europe. The Normans arrived as mercenaries in Southern Italy, torn apart by the struggles between the Lombards and the Byzantines, and obtained the county of Aversa from the duke of Naples, which became their first territorial base in the south. The progressive acquisitions were made official in 1059 by the Concordat of Melfi with which Robert Guiscard declared himself a vassal of pope Nicholas II in exchange for investiture as duke of Apulia (denomination of a territory that also included the current Basilicata and part of Campania), Calabria and Sicily (although the latter was still in Arab hands). In the following years Roger I, brother of Robert, conquered Sicily and received from the pope, in the absence of the papal legate, the apostolic legacy, that is the power to control the ecclesiastical institutions of the island, with the agreement to pay an annual income to the Roman

---

[19]Robert descended from a noble family of modest lineage ruling Hauteville-la-Guichard in Cotentin, a French peninsula stretching out over the English Channel. Guiscard's nickname has an uncertain etymology. Perhaps it is the combination of the Saxon terms *wis* and *hard*, meaning wise and strong; or it is simply the Italianization of the toponym of the place of origin, Guichard, which in French means fox, a symbol of cunning.

Church. Guiscard's nephew Richard II inherited the county, moved the capital Palermo (*Prima Sedes, Corona Regis et Regni Caput*), and in 1130, politically supported by antipope Anacletus II whom he had helped to be elected, was acclaimed king of Sicily.

Founded on an iron feudal structure centered on the figure of the sovereign, the reign of Sicily came to include, in addition to the south of Italy, also a large coastal strip in Tripolitania, Libya, and Algeria. Richard's reign was followed by those of his son William I, said the Wicked, and his nephew William II, the Good. In 1189 the latter died without heirs. He had indicated his successor in the person of his aunt Constance Hauteville who, according to Dante, was torn for the occasion from a convent by a paranymph pope (*Par.* III, 109–120). Things went differently, though. Breaking the oath of allegiance made to the dying sovereign, the Sicilian notables preferred her nephew Tancred, count of Lecce, chosen in anti-Swabian function; Constance, in fact, had married the son and heir of the Swabian emperor Frederick I Barbarossa. Before looking at the consequences of this action, let us return to the last years of Henry IV's reign.

2.7   Henry V of Franconia and Conrad of Hohenstaufen

With the death of Gregory VII, in 1085, a new period of great confusion had begun for the Roman pontificate. On the contrary, it seemed that the emperor had recovered all of his authority thanks to some lucky military campaigns against the indomitable Matilda in Italy and his own rebellious sons in Germany. For his part, continuing with determination along the political line drawn by his great predecessor, the new pope, Urban II, had tried in vain to bind the marquise with Guelph V of Bavaria in an anti-Germanic function. While the marriage was a disaster, it bound in Italy the name of Guelph to the cause of the pope, whose supporters will be called Guelphs.

Henry's relationship with Rome was then irremediably compromised, so much so that, calling the First Crusade, Urban did not bother in the least to address the emperor. The offence turned into a low blow in 1102 when his successor, pope Paschal II, formerly a Benedictine monk, issued him a new excommunication: a de-legitimation that four years later his son Henry V used to force his parent to abdicate. So miserably ended 50 years of reign during which the pope and the nascent communes in Central-Northern Italy acquired the awareness of being able to validly oppose imperial authority, presenting themselves, each one in his own capacity, as poles of a new world.

As we have seen, Henry V had ascended to the throne of Germany owing to an unintentional assist from the pontiff; but as soon as he took office, he resumed the anti-papal policy of his father, continuing to appoint bishops (who had become real cornerstones of an ever more distributed power). In 1110 the conflict led the *rex Romanorum* to a first descent to Italy with a large army to induce the pope to crown him emperor. The sovereign made a stop in Florence (the city of Dante, which thus appears for the first time in our story), then headed for Rome. The armed confrontation was avoided by an agreement reached with the pope. But the terms affected the interests of the German bishops, who rebelled. Not to displease them, Henry V was forced to take back his word. The issue could be solved only by force; so he arrested the pope, but a popular revolt forced him to leave Rome. An agreement was finally reached between the parties and the German king received his imperial crown in St. Peter's.

Barely a year passed and this time it was the pope who retraced his steps. While Henry was seriously committed to containing turbulence on the domestic front, he was even excommunicated by the archbishop of Milan. It was a stab in the back that forced him to return to Italy where he was crowned emperor by the papal legate, given that the pope had fled from Rome. The following year he had to return again to the city, called by the opponents of the new pontiff, whom he deposed by appointing his own antipope. History was repeating itself according to a script already seen, with a second excommunication of Henry and with serious damages to both sides. Thus, in 1122 Henry V and pope Callistus II reached the concordat stipulated in the town of Worms, in Rhineland-Palatinate, with which the dual temporal and spiritual function was recognized to the bishop. In Germany the emperor was present at the election of the bishops with the right of veto, while in the kingdoms of Burgundy and Italy he was not present and only six months later he granted the investiture of the benefits. It was the apotheosis of the political conception known as the Two Suns, which saw the papal and imperial authority of equal dignity, but referred to different areas, including territorial ones: a political theory of which Dante, although Guelph, would have become spokesman in the *De Monarchia* and in the *Commedia* (*Pur.* XVI, 98–99):

> ... 'l pastor che procede,
> rugumar può, ma non ha l'unghie fesse

"...the shepherd that leads can chew the cad but does not have cloven hooves".[20]

Shortly thereafter, Henry V died leaving no male heirs. Although he had indicated as his successor to the title of *rex Romanorum* his nephew Frederick Hohenstaufen,[21] duke of Swabia, the choice of the German electors fell on the name of Lothair II, duke of Saxony, who had been able to skillfully navigate the long struggle between the Salian and the Saxon dynasties. With him, the center of gravity of imperial power again passed from Swabia to Northern Germany. Two years later Frederick Hohenstaufen, who had to swallow the election of Lothair, rebelled and had his younger brother Conrad elected king of the Romans. Promptly descended into Italy, Conrad was crowned king of Italy in Monza.

Meanwhile in Rome, due to the conflict between two powerful families who controlled the city, a real schism of the papacy had been created. The conclave called to elect the pontiff had expressed not one but two popes, Anacletus II and Innocent II, champions of the opposing factions. Not a pope and an antipope, therefore, but two pontiffs, each of whom considering himself entitled to reign. Anacletus, who had the support of the Normans, seemed to gain the upper hand over his rival. But Innocent, supported by the authoritative voice of the French Bernard de Fontaine (Bernardus Claravallensis), a Cistercian monk and theologian, skillfully managed to be elected on the death of Anacletus II in 1138 and to return to Rome. He had in his favor the kings of France and England, to whom was also added Lothair who had taken up arms in Italy where he received the coronation as emperor. Upon his return to Germany, this latter had died and the eternal rival Conrad of Hohenstaufen had been elected king of the Romans. His actions had no consequences for Italy, except for the designation of his successor, his nephew. Frederick I, known as Barbarossa.[22] Conrad consumed his cards by choosing to participate in the Second Crusade promoted by

---

[20]Apparently sibylline verses meaning that the pope who guides the flock (*il pastore che procede*) can ruminate (*rugumar*), that is he can interpret the sacred texts, but he has not split (*fesse*) nails: he does not distinguish and keep separate, as he should, the temporal authority from the spiritual one.

[21]The family of Hohenstaufen owned the castle of Waiblingen in Baden-Württemberg. The name of this residence, italianized in Ghibellines, would be used starting from 1242, to indicate the supporters of the imperial party in antagonism with the papal one.

[22]Red Beard: a nickname given to him by the Northern Italian cities he tried to rule. In Germany, he was similarly named *Kaiser Rotbart*.

Bernard of Clairvaux, distinguishing himself for his useless cruelty and miserably losing his army, defeated by the Turks in Anatolia.

In 1152, the thirty-year-old Frederick I of Swabia was unanimously acclaimed king of Germany and solemnly crowned in Aachen. Strong, ambitious, shrewd, tyrannical, and at the same time cynical manipulator of men, but also, in the imagination of the German nation, enlightened sovereign and loyal and courageous knight, he showed from the beginning the will to impose the imperial power *urbi et orbi*. Leaving aside the internal conflicts in the Germanic world, which at his appointment were undermining the integrity of the German nation and the figure of the emperor himself, and that Frederick was able to control with determination but also cunning and foresight through concessions and marriages, we will focus on the initiatives to realize his dream of a universal civil power. That would include imperial supremacy over the papacy and control over all German and Italian territories, including both Norman-held Sicily and Italian communes. The latter, taking advantage of various circumstances, had in fact constituted themselves in cohesive units relatively independent from the authority of the emperor's feudal lords, to whom they bestowed just loyalty and the right of gabelle.

2.8   The birth of the communes in Italy

Around the year 1000,[23] in Italy there was an explosion of new energies, favored by many factors that we will try to analyze in brief. First of all a demographic explosion that continued for at least three centuries, accompanied by an improvement in the climate and by a wider availability of food resources. The lands were better cultivated, with increasingly refined procedures and tools (for example, the three-year rotation of crops and new types of plow). The greater prosperity caused the demand for goods to flare

---

[23]The story that the Europeans were terrified by the approach of the year 1000, thought to mark the end of the world, is not reflected in any document of the time. It is a legend built in the XIX century, founded on a prophecy of St. John's *Apocalypse* fixing Satan's imprisonment in a thousand years, and on statements attributed in the apocryphal Gospels to Jesus Christ. In short, "a thousand and no more than a thousand" is part of the general misrepresentation of an era that is anything but dark: a derogatory attitude that began with Giovanni Petrarca and his regret of the lost ancient grandeur, and became popular with the Enlightenment. Consider, for example, the adjective Gothic, equivalent to barbaric, associated with the style of the great architectural achievements that arose in Europe from the mid-XII century.

up, with the consequent flourishing of a new social class of individuals committed to enriching themselves by promoting manufacturing activities and the circulation of goods and money; the merchants (from the Latin *mercari*, to trade). A different typology from the classic ones of prelates, soldiers, and peasants (*oratores, bellatores, laboratores*).

To broaden the horizons of their trades to new markets and new products, the merchants resumed the old roads and opened new ones, including the ancient silk road in the East. Their needs motivated technological innovations (for example, the compass), the birth of professions and skills related to the management and exploitation of money (letters of credit, banks, accounting), the resurgence of large fairs, and a greater need for schooling. Owing to their trades, some families accumulated huge wealth that made them a class yet distinct from the ancient aristocracy, but ready and prepared to replace the old powers by renewing blood and rules. Pragmatic and not very ideological like those who have to pay attention first to their business, they had their nerve centers in the cities which, under the impulse of manufacturing and commercial activities, became populous centers of economic power, real nodes among the trade routes; and as it has always been in history and will always be, well-being led to a flourishing of culture, art, architecture, and the development of university education. The new cultural climate favored the rediscovery of the ancient splendor and dormant pride, a new desire to feel in control of themselves and their things. From the point of view of economic thought, the mercantile elites resisted the attacks of the pauperistic movements which were also very successful in this period.

In short, the spread of the mercantile system would have contributed, between the XII and XV centuries, to eliminate many of the natural, cultural, moral, religious, ideological, and intellectual barriers, and to shape a new society and a new geography in Europe. Even Purgatory, conceived as a physical parking space for the souls of those who, while dead in the grace of God, are not yet ready for the Heaven, and on which Dante would have built an entire canticle of the *Commedia*, was introduced in the Catholic doctrine in the second half of the XII century with the function of reassuring the professional sinners, that is those merchants whom Jesus had expelled from the Temple.

From a political and administrative point of view, a strongly autonomous movement spreads, in contrast with the universalistic principles of the papacy and the empire (in which also Dante believed, with the caveat that they should stay separated). Thus the first communes flourished in Northern-Central Italy, in cities where the social composition was very heterogeneous (unlike the towns of Flanders where the merchant classes made up a large part of the population). Inhabited by the patronages of the vassal-beneficiary bishops, by aristocracies of ancient origin connected to the possession of the land, and by the emerging classes of notaries, jurists, and merchants, the cities of Italy had not a common profile. Almost everywhere, however, in the early stages the communes were administered by the *milites*; but the tumultuous growth of the society caused soon, especially in some cities with a strong mercantile vocation, the emerging of other social classes. Among the most flourishing communes there were, for example, Milan at the center of the territories that had already belonged to the kingdom of the Lombards, and in Tuscany, while less well organized, the cities of Lucca, Siena, and Florence, as well as the so-called maritime republics, Venice, Genoa, and Pisa, which more than others used the sea as a trade route. We must not think of these microcosms as renewed niches of Athenian democracy. They were rather oligarchies where few shared the power following rules designed to maintain the equilibrium among the various forces. The government action was entrusted to the figures of some consuls (typically from 2 to 20), renewed with considerable frequency, and later to a sole chief magistrate (*podestà*), holder of the highest civil office: an equilibrium figure who ended up to be a foreigner so as not to favor anyone. And while the Holy Roman Empire continued to shatter into an archipelago of Italian and German fiefdoms, less and less tied to imperial power, and in France, Sicily, Hungary, England, and Spain the first national states consolidated, some communes, expanding their sphere of influence on the countryside, extended to create what at the end of the Middle Ages became regional states.

2.9   The Italian politics of Frederick I Barbarossa

Heir to the Ghibellines of Franconia and at the same time related to the Guelphs of Swabia through his mother, Frederick I Barbarossa collected in him the two main families of feudal Germany, for over a century in bloody competition for power. He was therefore in the ideal condition to start the construction of a Germanic national state that he wanted governed by just laws, equal for all. But Frederick pursued an even higher ideal, the realization of a universal empire. It would have been this extreme ambition to frustrate his efforts, despite the favorable starting conditions.

After he had pacified Germany with some generous concessions to his potential rivals, in 1154 the king turned to Italy with the intent of reducing the demands of the pontiff, securing control over the rebel communes in the north and center of the peninsula and take possession, in the south, of the kingdom of Sicily, in the hands of the dynamic Norman dynasty: a dream already pursued without success by almost all of his predecessors since the progenitor Otto I. As already happened many times before him, the opportunity came as a reckless request. Some Lombard communes asked him to act as arbiter in the disputes that broke out in a context of effervescent growth and excessive fragmentation, where every party tried to assert his own "particular",[24] often regardless of the consequences. A situation that Dante would have summarized in the famous invective of canto VI of *Purgatory*, almost a vent at the sight of the embrace between his guide Virgil and the Mantuan poet Sordel (vv. 76–120) and at the same time *j'accuse* towords a corrupt and rapacious Church, an emperor who is failing in his duty, and even Christ (Jupiter), who has averted the gaze from Dante's country:

> *Ahi serva Italia, di dolore ostello,*
> *nave sanza nocchiere in gran tempesta,*
> *non donna di province, ma bordello!* ...
>   *e ora in te non stanno sanza guerra*
> *li vivi tuoi, e l'un l'altro si rode*
> *di quei ch'un muro e una fossa serra.* ...
>   *Che val perché ti racconciasse il freno*
> *Iustiniano, se la sella è vota?*
> *Sanz'esso fora la vergogna meno.*
>   *Ahi gente che dovresti esser devota,*
> *e lasciar seder Cesare in la sella,*
> *se bene intendi ciò che Dio ti nota,*
>   *guarda come esta fiera è fatta fella*
> *per non esser corretta da li sproni,*
> *poi che ponesti mano a la predella.* ...
>   *Vieni a veder la tua Roma che piagne*
> *vedova e sola, e dì e notte chiama:*
> *"Cesare mio, perché non m'accompagne?".* ...
>   *E se licito m'è, o sommo Giove*
> *che fosti in terra per noi crucifisso,*
> *son li giusti occhi tuoi rivolti altrove?*

---

[24]Expression coined by Francesco Guicciardini, Florentine writer, historian, and politician of the XVI century, in antithesis to the concept of universal.

"Ah, slavish Italy, dwelling of grief, ship without a pilot in a great storm, not a ruler of provinces but a whore! [...] and now in you the living are not without war, and of those whom one wall and one moat lock in, each gnaws at the other! [...] What does it profit that Justinian fitted you with the bridle, if the saddle is empty? Without the bridle the shame would be less. Ah, people who should be devoted and permit Caesar to sit in the saddle, if you attend to God's words to you, see how this beast has become savage, not being governed by the spurs, ever since you seized the reins. [...] Come and see your Rome, which weeps widowed and alone, and day and night calls out: 'My Caesar, why do you not keep me company?' [...] And if it is permitted me, O highest Jove, who were crucified on Earth for us, are your just eyes turned elsewhere?"

For his part the pope, while not liking Barbarossa and fearing his ambitions, was waiting for him in the Eternal City for a consecration to emperor in exchange for the head of Arnold of Brescia,[25] the pupil of Abelard who was trying to reform the Church in an anti-papal function, and of a protection against the aggressiveness of the Normans, who from paladins of the bishop of Rome had now turned into rapacious aggressors. Thus, in the autumn of 1154, Frederick embarked on his first *Romzug*, during which he was crowned as king of Italy in Pavia, then showed his powerful muscles to some rebel cities. Finally he reached Rome where he received the most coveted legitimacy, the consecration as emperor by the pope. But, failing to tame the riotous Romans, he returned to Germany in time to extinguish some revolts, leaving everyone unhappy: the northern communes, that would have later created a Lombard League to oppose him, and the pope, who was left alone to defend himself from the Normans.

Frederick returned a second time to Italy in 1158 to curtail the hegemonic ambitions of Milan and reaffirm the rights of the emperor (*iura regalia*) over municipalities then unwilling to accept interference in their political, administrative, and fiscal life. The armed confrontation was inevitable. In the meantime, Frederick had moved to Central Italy to conquer the Tuscan fiefs formerly belonging to Matilda of Tuscany. Upon her death in 1115, this ardent supporter of the Church had left them as

---

[25] Augustinian canon, Arnold of Brescia headed an anticlerical movement that demanded the return of the Church to evangelical poverty, with the disavowal of all temporal power and the transformation of Rome into a free commune, to the point of preaching the priesthood of every Christian. Excommunicated, and forced to flee, he was captured by Frederick's troops and handed over to the pope who had him hanged. His body was then burned to prevent relics from being collected.

an inheritance to the Church. Pope Adrian IV, who even had crowned Barbarossa, was furious and began to plot with the hated Normans against the emperor, but soon died. Frederick then thought that the time had come to deliver the *coup de grace* to his antagonists. In opposition to the new pope, the Sienese Alexander III, elected by the Romans, he managed to nominate an antipope, Victor IV, choosing a member of the Crescenzi family. Then he turned towards the Milanese rebels, besieged their city, and, upon the surrender of the municipal militias, razed it to the ground. Partisan of the model of the Two Suns, Dante will not condemn the German antichrist whom he even places in the *Purgatory*, recalling with veiled satisfaction the severe punishment inflicted by the *"good Barbarossa"* on the Lombard commune, guilty of treason (*Pur.* XVIII, 120):

> *di cui dolente ancor Milan ragiona.*

"of whom Milan still speaks with grief."

Alexander III, however, did not give up. Averted an alliance between Frederick and the king of France, he managed to resettle himself in St. Peter's from where the emperor, by defeating the Romans in 1167, drove him back again into the arms of the Normans. Having installed a new antipope, Barbarossa had himself crowned emperor for the second time. Then he was forced into a disastrous retreat to Germany by the explosion of a violent epidemic which had begun to decimate his troops. Meanwhile, his overbearing arrogance had ended up cementing an alliance of the main municipalities of Northern Italy. Under the benevolent eyes of Venice, Rome, and Constantinople, Milan, Lodi, Ferrara, Piacenza, and Parma temporarily overcame diversities, egoisms, and frictions and signed pacts[26] to counter Barbarossa's hegemony by helping each other. Their intent was to demand from Frederick the recognition of the autonomies granted to them by Henry V.

For the emperor, things were no better in Germany. The heavy war expenses and the uninterrupted conflict with the papacy displeased both the bishops and the great vassals, who would have preferred to use those resources to expand the empire to the north-east. But Frederick did not want to hear reasons or accept advice. Stubbornly, he went down to Italy for the fifth time, determined to make a clean sweep of Milan, which in the meantime had been rebuilt, but he was soundly beaten by the Lombard

---

[26]The tradition born in the XVI century indicates the place of the oath in the abbey of Pontida, near Bergamo, and the date as April 7, 1167.

League in the battle of Legnano, about 20 kilometers north-west of Milan, on May 29, 1176. Nor could he risk a revenge, because the residual forces served him home to tame once and for all the rebellion of the Guelph Henry the Lion. He had to yield on both the front of the communes, that acquired the coveted autonomy, and of the supremacy of his on the pope's Sun.

Defeated but indomitable, the old emperor tried one last move to realize the dream that had never come true before: the conquer of the south of Italy. Lacking the strength, he used cunning, tying his young son, the future Henry VI, to a promise of marriage with the princess Constance, 15 years older than him but, as we have seen, designated heir of the Norman kingdom of Sicily. Three years later, like a hero of a Nibelungs saga, he left for Jerusalem to give back to Christianity the Holy City that had fallen into the hands of Salah ad-Din, drowning during the journey in 1190. He left Italy stronger and more aware of her possibilities, an indomitable pope well determined to expand his influence and his temporal power, but also a son who harbored his own ambitions.

2.10 The crisis of the empire: Henry VI and Otto IV

Even before being elected king of Germany, owing to the marriage with Constance of Sicily celebrated in Monza in 1186, the twenty-one-year-old Henry found himself in the position of heir to a kingdom coveted by all the peoples of the cold north. Three years later, on the premature death of the Norman king William II the Good, he was in the position of claiming the crown. But the Sicilian barons and prelates, heartened by the fact that the absence of Barbarossa, engaged in the Holy Land, forced Henry to preside over Germany, gave their preference to Tancred of Lecce, cousin of the late king and therefore nephew of Constance:[27] a reckless but cultured man, valiant and courageous warrior.

The following year, however, Frederick I died and Henry, having taken possession of the throne of Germany and settled again the dispute with Henry the Lion, king of Bavaria, cousin of his father Frederick and rocky champion of the Guelph side, descended to Italy. He was first crowned emperor by the weak pope Celestine III, raised to the throne of Peter when he was already eighty-five. Then he moved his army south with the intention of getting rid of Tancred and securing the kingdom of Sicily as his

---

[27]Tancred was the illegitimate son of Richard III of Apulia, in turn the eldest son of Roger II the Norman, king of Sicily, and brother of Constance.

personal property, released from the empire. But, because of the strenuous resistance of Naples and of a deadly epidemic, he had to give up, after having even suffered the temporary capture of his wife, who had stopped in Salerno to be treated by the doctors of that famous medical school. It was a move perhaps necessary but fraught with consequences for a man of unusual ambition. His retreat, in fact, persuaded his enemies (the pope, the heir of Henry the Lion, and the King of England Richard the Lionheart) to recognize the legitimacy of Tancred.

Within a year, however, thanks to a fluke (the sudden death of Tancred) and to a skillful move to the limits of piracy (the capture and consequent neutralization of Richard the Lionheart, who had been intercepted by the duke of Austria, an ally of Henry, while the English sovereign was returning from the Third Crusade), the emperor was able to turn events in his favor. He began with taking over the kingdom of Sicily, drowning all opposition in blood. Then he placed family and faithful friends in the main fiefdoms of Central and Northern Italy, while keeping the German feudal lords in control. Seized by imperial bulimia, he began to overflow on the other side of the Mediterranean Sea, into North Africa and even in Constantinople, meeting less and less resistance as his power grew. Even the pope then felt crushed by the excessive power of Henry, with a State of the Church reduced to a handkerchief around Rome. However, in spite of so much power, Henry was unable to transform the empire into a hereditary monarchy. Indeed, when in 1197 he suddenly died in Messina, perhaps poisoned by his wife, it was precisely his arrogance combined with the ruthless methods of government that determined the disintegration of his vast empire. He left the kingdom of Sicily much poorer and more disordered than he had found it, an even more aggressive pope, and a temporarily cohesive Italy of the municipalities. He also left behind a three-year-old son, the future Frederick II, *Stupor mundi* (wonder of the world), exalted by Dante[28] in the *Commedia*, the *Convivio* and *De vulgari eloquentia* as a promoter of philosophy and of the first Italian poetic school, but then placed him in Hell among those "*che l'anima col corpo morta fanno*" (*Inf.* X, 15) "who make the sould die with the body", because of his alleged heresy.

---

[28] "*Quest'è la luce de la gran Costanza / Che del secondo vento di Soave / Generò 'l terzo e l'ultima possanza*", (This is the great Constance who by the second wind of Swabia generated its third and last power; *Par.* III, 118–20); an exaltation of the Swabian dynasty by a Guelph of Ghibelline faith (White Guelph).

2.11   Emperor Frederick II of Swabia and king Manfred

Frederick inherited the kingdom of Sicily, but his very young age exposed him to serious dangers. For this reason his mother Constance, before dying herself in 1198, had arranged for him to be entrusted to the tutelage of pope Innocent III. Actually, the leadership of the kingdom and the education of the child both fell into the hands of the German and Norman feudal lords. In 1209, now of age, Frederick took Constance of Aragon as his wife. The marriage had been arranged by the pope who intended to use the young man to implement his temporal policy. Innocent had already entered heavily in Frederick's life on the occasion of the struggle for the succession to the German throne of his father. Let's see how and why this affects so much the history of Italy and of Florence itself.

After the death of Henry VI, in 1198 the majority of the German princes had elected as *rex Romanorum* his younger brother Philip, duke of Swabia; again a Ghibelline. A minority of the electors, however, had opted for a Guelph anti-king in the person of Otto of Brunswick, third son of the rebel Henry the Lion and of an English princess (a not negligible detail because it marks the entry into the game of England and, indirectly, of France). Trying to avoid the encirclement of his temporal domains by the Ghibellines, pope Innocent had chosen to side with Otto IV, even crowning him emperor in exchange for the promise of support for his expansionist policy. But in 1210, feeling perfectly in the saddle, Otto threw off his mask and began to behave like a master even towards the pontiff, demanding from Frederick the delivery of the Sicilian kingdom. At this point Innocent unsheathed the secret weapon of the popes: the excommunication. It was a breathtaking shot. The German princes, who had supported Otto, were now abandoning the Guelph emperor, accused of pursuing his obsession with hegemonic power at the expense of defending Germany from the aggressions of the Danes. Put on the ropes, Otto tried everything, including a plea to the pope for forgiveness. It was too late. Innocent III, who had already been declared by a Lateran Council the superiority of the Church over any other secular power, as the sole custodian of Grace and exclusive mediator between God and men, had already struck his fatal blow. He had deposed Otto and appointed as emperor the young Frederick, who, without wasting time, had rushed to Cologne to be crowned *rex Romanorum*. His rival, the Guelph Otto, tried to make up for himself by allying to the English crown, but was defeated in the battle of Bouvines, in 1214, and definitively neutralized. It should be noted that, thanks to the pope, the Ghibelline party was winning hands down.

Having settled the German question not without difficulty, Frederick triumphantly returned to Italy. In 1220, in exchange for the solemn promise to participate in the liberation of the Holy Land (Sixth Crusade), he was crowned emperor in St. Peter's by pope Honorius III. His predecessor, Innocent III, had died four years earlier after two decades of pontificate during which he had relaunched the power of Rome *urbi et orbi*. Now master of the world, Frederick II went down to his Sicily where he started a reform program conducted in an authoritarian way, which yet gave those warm lands a moment of absolute splendor in the European panorama. Among the skilled officials who assisted him in this drawing, there was the Capuan scholar Pier delle Vigne, whom Dante meets in the seventh circle of Hell, in the forest of suicides (*Inf.* XIII, 58–72). Transformed into a plant, Pier tells the poet of his close relationship with the emperor and of the suicide committed to save his honor (*"col morir fuggir disdegno"*, "by death to flee disdain") soiled by envy (*"meretrice"*, "prostitute"):

> *Io son colui che tenni ambo le chiavi*
> *del cor di Federigo, e che le volsi,*
> *serrando e diserrando, sì soavi,*
> *   che dal secreto suo quasi ogn'uom tolsi:*
> *fede portai al glorioso offizio,*
> *tanto ch'i' ne perde' li sonni e 'polsi.*
> *   La meretrice che mai da l'ospizio*
> *di Cesare non torse li occhi putti,*
> *morte comune e de le corti vizio,*
> *   infiammò contra me li animi tutti;*
> *e li 'nfiammati infiammar sì Augusto,*
> *che 'lieti onor tornaro in tristi lutti.*
> *   L'animo mio, per disdegnoso gusto,*
> *credendo col morir fuggir disdegno,*
> *ingiusto fece me contra me giusto.*

"I am he who held both the keys to the heart of Frederick and turned them, locking and unlocking, so gently that I excluded almost everyone else from his intimacy; I kept faith with my glorious office, so much that because of it I lost sleep and vigor. The whore who never turns her sluttish eyes away from Caesar's dwelling, the common death and vice of courts, inflamed against me all spirits; and those inflamed Augustus so that my bright honors turned to sad mourning. My spirit at the taste of disdain, made me unjust against my just self."

The promise of a crusade hung: the emperor did not want to give free rein to the pope's intrigues. After yet another postponement, Gregory IX sent him a heavy warning: a first excommunication. So Frederick, not at all enthusiastic, in 1228 had to leave his favorite falcon hunting and embark for the Holy Land where, thanks to his marriage to Yolande of Brienne,[29] he had already assumed the title of king of Jerusalem. This Sixth Crusade was not a bloody expedition. The emperor made an agreement with the Muslim sultan, managing to avoid an armed confrontation. Jerusalem was handed over to the Christians peacefully but deprived of the walls and therefore indefensible.

Returning to Italy, Frederick had to face the hostility of the pope and the rebellion of the northern municipalities. Winds of revolt were also blowing in Germany, fueled by one of his sons. The emperor came to the head of it, then he turned against the Lombard League giving it a resounding and humiliating defeat. Pope Gregory IX, however, did not give up and, after yet another quarrel over a territorial matter, issued another excommunication without being able to have Frederick deposed. But it was only a matter of time. Gregory IX's successor, Innocent IV, although less energetic, renewed the excommunication and declared him lapsed, thus rekindling the fuse of the anti-imperial revolt in Italy and Germany. Aware of his weakness, Frederick tried in vain to negotiate with the pope and in 1250, heartbroken by the disappearance of his favorite son, he suddenly died, leaving the regency of Italy to Manfred, his illegitimate son.[30] This was a temporary solution, as Manfred had to keep the throne for the true heir, his brother Conrad IV, who in the meantime was putting down the claims of an usurper in Germany.

With Frederick an extraordinary character disappeared from the history of Europe, who had had the merit of promoting the meeting of three civilizations, Latin, Greek, and Arab: a protector of arts and sciences, a man of letters himself, founder of universities, often enlightened administrator. He was the champion of the courteous civilization, the world of "*donne e' cavalier, li affanni e li agi / che ne 'nvogliava amore e cortesia*" (*Pur.* XIV,

---

[29]Yolande was the daughter of John of Brienne, king of Jerusalem and Latin emperor of Constantinople. In 1099, during the First Crusade, Jerusalem had been conquered by the army led by Godfrey of Bouillon. The Holy City remained in Christian hands until the reconquest by Salah ad-Din in 1187. The capital of the Christian kingdom was then transferred to Acre where it remained until 1291. The sovereigns, belonging to the great families of the European nobility, almost never resided in the Holy Land.

[30]Perhaps Manfred was legitimized by his father with a marriage on the verge of death.

109–110), "the ladies and the knights, the labors and the leisures that love and courtesy made us desire".

In character and qualities, Manfred resembled his father. Dante describes him as follows: *"biondo, bello e di gentile aspetto, / ma l'un de' cigli un colpo avea diviso"*, "blond and handsome and of noble appearance, but a sward-blow had divided one of his brows", to underline his bellicosity with the sword wound (*Pur.* III, 107–108). While acting on behalf of his half-brother Conrad IV, he initially managed to successfully stand up to Innocent IV who claimed to treat him as one of his vassals. But in 1254 Conrad died leaving a younger son, Conradin, under the tutelage of the pontiff. Deprived of all help and even excommunicated, Manfred had to bow his head and, to be forgiven, accept the papal occupation of some counties of the kingdom of Sicily. But when he had reconstituted an army, he marched against the pope, defeating him. Even Rome became Ghibelline and the pope was forced to transfer the papal see to Viterbo, north of Rome. At the height of his success, having spread the rumor that his nephew Conradin was dead, Manfred had himself appointed king of Sicily and crowned in Palermo.

The papal talent scouting in search of an antidote against Manfred became successful in 1265. Charles I of Anjou, brother of King Louis IX of France and "landless",[31] came to Italy with a powerful army and, once in Rome, was crowned king of Sicily by Clement IV. The Italian Ghibellines disbanded and the Angevin troops had their way cleared to the south. In Benevento, in the Apennine hinterland of Campania, they clashed with Manfred's army on which they inflicted a resounding defeat. It was February 26, 1266. The king himself lost his life fighting with desperate valor, surrounded by his Sicilian and Saracen militias. Even the Guelph Dante Alighieri will remember him in the *Commedia* with respect, placing him in the *Antipurgatory* despite repeated excommunications, among those who repented at the end of their life and were welcomed by the *"infinite goodness"*. Two years later, the Angevin repeated himself by defeating in Tagliacozzo, about 70 km east of Rome, the last Swabian king, Conradin, who for prudence was beheaded together with the most prominent nobles who had accompanied him in battle.

---

[31] Actually he was Count of Provence and Forcalquier.

The great poet Dante was born in Florence 8 months before the battle in Benevento; and it is his city that we will deal with in the last sections of this introductory chapter.

2.12 *"Florence within the ancient circle"* (*Par.* XV, 97)

Florentia was founded by the Etruscans at the point of the Arno river valley below Visul (Fiesole), where the river was probably narrower and easier to ford and the land less marshy.[32] In the II century AD it became a Roman municipality. Disputed during the Byzantine–Gothic war because it was along the communication route between Rome and the north-east, in 570 Florence fell into the hands of the Lombards. It was precisely this strategic position, which would later make the fortune of the city, to cause its temporary decline. For travelling between north and south, the Lombards favored the western path through Lucca, safer than the central itinerary through Bologna–Pistoia–Florence, too exposed to the incursions of the Byzantins.

In 774 Florence became part of the administrative system of the Carolingian kingdom. About a century later it was equipped with walls, as well as new military, civil, and religious buildings. The progressive downsizing of the role of Lucca and an active participation in the clash between the papacy and the empire contributed to increasing the importance and wealth of the city, which in 1055 was chosen for a council held in the presence of Emperor Henry III. Florence was also the only city in Tuscany to support the cause of the marquise Matilda of Tuscany when she was deposed in 1081 after the defeat inflicted by Henry IV. A gesture of loyalty that cost Florence a siege by the imperials, passed owing to the "new walls" (the Matildine walls, in fact) mentioned by Cacciaguida, Dante's ancestor, in canto XV of *Paradise*.

In these years, thanks to the marquise's concessions, the conditions for the transformation into a commune with hegemonic ambitions matured. It formally happened in 1115, with the death of Matilde herself. The vacuum of power which followed meant a strengthening of the control of the eminent Florentine families over the countryside. When, ten years later, Henry V also died, the Florentines started a policy of expansion against the nearest towns: among these Fiesole, which was destroyed and its inhabitants forced

---

[32] According to some authors, Florentia is the Latin adaptation of the Etruscan word Birent or Birenz which means swampy place. The hypothesis is contested by those who believe that it is a typical augural toponym derived from the verb *florere*, that is to flourish.

to move to Florence. And in the year 1138 its consuls formed a league with other Tuscan communes to counter the imperial claims.

The dominant classes were the aristocrats, grouped into factions and major holders of power, and the people, made up of the class of notaries, lawyers, merchants, and artisans. Taking advantage of the strategic position and of the alliance with Pisa, which guaranteed both towns the exploitation of the Arno as a waterway, Florence had in fact started a dense network of exchanges with an ever wider range, based on the procurement of artifacts, in particular cloths from Northern Europe, and on their subsequent sale after an improvement process. As we have already pointed out, this activity led to the rapid accumulation of large amounts of wealth, to be invested in land, in small artisan businesses, and in money itself: anyone with liquidity did not hesitate to lend it at interest. Credit was the lever of the city's growing prosperity but also the origin of its ills, as Dante claims through Folquet de Marselha (*Par.* IX, 127–142):

> *La tua città, che di colui è pianta*
> *che pria volse le spalle al suo fattore*
> *e di cui è la 'nvidia tanto pianta,*
> > *produce e spande il maladetto fiore*
> *c'ha disviate le pecore e li agni,*
> *però che fatto ha lupo del pastore.*
> > *Per questo l'Evangelio e i dottor magni*
> *son derelitti, e solo ai Decretali*
> *si studia, sì che pare a' lor vivagni.*
> > *A questo intende il papa e' cardinali;*
> *non vanno i lor pensieri a Nazarette,*
> *là dove Gabriello aperse l'ali.*
> > *Ma Vaticano e l'altre parti elette*
> *di Roma che son state cimitero*
> *a la milizia che Pietro seguette,*
> > *tosto libere fien de l'avoltero.*

"Your city, planted by him who first turned his back on his Maker and whose envy is so much bewailed, brings forth and spreads the cursed flower that leads the sheep and lambs astray, since it has made a wolf of the shepherd. For this the Gospel and the great Doctors are forgotten, and only the Decretals are studied, as their margins show. To this the pope and cardinals attend; their thoughts go not to Nazareth, where Gabriel opened his wings. But Vatican and the other noble parts of Rome, cemetery of the army that followed Peter, will soon be freed from such adultery."

The overwhelming demographic development of Florence, with the urbanization of people from the countryside, had only a brief setback when, in 1185, Frederick Barbarossa tried to restore the imperial control over the town. The city was growing in size, so much so that it required a new wall system to defend against external enemies. Imposing on the skyline together with the bell towers of the churches, there were dozens of toll towers erected to mark the wealth and influence of the owners, a sign of permanent tensions between the richest or enriched families. At the end of the XII century, the city had expanded: the walls went beyond the Arno, encircling some villages inside. A main role in the latest reorganization of the medieval city would later be played also by the new religious orders, Franciscan, Dominican, Augustinian, Carmelite: creative presences and critical consciences but also sources of turbulence.

Not unlike the other communes, the government was initially entrusted to consuls, figures inspired by Roman magistrates with similar powers. Elected among the families of the city elites, they only held office for a few months, to avoid any accumulation of power. An institutional system that was only partially democratic because, among other things, Florence was still subject to imperial power. In any case, the struggles to access the consulate split the knightly class to such an extent at the end of the XII century that it was necessary to change the institutional structure: from 1197 a *podestà* with executive power functions was elected in place of the consuls.[33]

There was another important social component in the city. With the development of trade and entrepreneurial activities, the corporations of the major and minor Arts took hold. Their members were active not only on the economic side, but also on that of city defense: they formed the infantry of the municipal army.[34]) It is this component, known as the People, which

---

[33]In 1197 the *podestà* is Florentine. Only ten years later, in 1207, the habit was introduced to enrol a foreigner, following the current practice in the center and north of Italy. The *podestà*, who remained in office for a year, had the executive power, but was supported by a series of local councils that in fact widened the possibility of political participation. It is at this point that the families of merchants, entrepreneurs, etc., entered the public scene, asking for their own representative, the captain of the People (also a foreigner).

[34]Among the major Arts stands that of Calimala, which brought together the great wool merchants and which was originally based in the Cavalcanti house, on the ground floor of one of the family towers. The etymology of Calimala is uncertain: it could be from the Latin *callis mala* or *callis maia*, meaning bad road or major road (*cardus*), or from the Greek *kalos mallos*, beautiful wool.

found political space with the crisis of the consulate: the *podestà* was in fact flanked by two councils (general and special) to which the most prominent exponents of the Florentine merchant and entrepreneurial class had access (between 1244 and 1248, the People will have their own representative, the captain: he will govern alongside the *podestà*).

At the beginning of the XIII century, therefore, Florence was a lively city from the economic, social, and political point of view. As such it was not alien to the struggles between the two great powers of the time, the emperor and the pope; and the knightly class, which had never known true unity, returned to tear itself apart. The continuous struggles between factions, often disguised as more or less sincere supporters for one of the two parties fighting for power in the Western world, the Guelphs, defenders of the superior authority of the Church, and the Ghibellines, supporters of imperial supremacy, produced a conflict which began around 1215. It developed with alternating events, between feuds and peacemaking marriages, until the arrival in Florence, in 1248, of an imperial vicar Frederick of Antioch, son of Frederick II, who decidedly tipped the scales in favor of the Ghibellines. But, with the death of the emperor, in 1251, the Guelphs returned to power and took revenge on their enemies by exiling them. Among them the Florentine Farinata degli Uberti, who found a way out in Siena and whom we will meet again shortly.

These were difficult times but of extraordinary economic growth. The city was governed by the People: having restored the figure of the captain of People, with military command functions, Florence adopted a new council (of the 12 Elderly) and nominally wanted to disengage from belonging to the Guelph or Ghibelline party (the government was still made of businessmen, interested in their own gain rather than in ideological issues). It was a period of military successes and of great economic flourishing: the gold florin, minted for the first time in 1252, owing to the growing financial power of Florence would have become the preferred exchange currency in Europe, with dignity equal to that of the Venetian duchy. Three years later the Bargello town hall was built.

Meanwhile, the contrast with the nearby commune of Siena, where the Ghibelline party prevailed, was intensifying. In reality, the reasons for the clash were economic rather than ideological. Both Tuscan cities, taking advantage of their diversity (Siena was located on the Via Francigena, the pilgrim route to Rome, while Florence compensated for less valid land connections with the availability of a waterway), had reached a considerable economic power thanks to trade; and both competed to secure the pope's role as debt collectors and bankers.

In 1260 the Florentines, convinced to be in clear advantage, attacked the Sienese army. The clash, full of consequences for the history of Italy and perhaps of Europe itself, took place on September 4, 1260, in the plain of Montaperti, near a village bathed by the Arbia stream and a few kilometers away from Siena. The largest army was that of the Florentine Guelphs and their allies, all supporters of the pope. The Sienese Ghibellines, far out-numbered, partisans of imperial sovereignty, and opposed to the invasion of Rome in their affairs, confronted them in arms; they were flanked by a thousand German knights sent to their aid by Manfred as a result of an alliance stipulated the year before. These are war professionals, who com-pensated with a formidable training the lack of enthusiasm characteristic of those who do not fight for themselves. The outcome of the bloody battle, which involved a total of at least 55,000 soldiers, a huge number for those times, had been predicted by an astrologer, Guido Bonatti, who earned fame and glory for this prophecy. The Sienese won, exploiting the harsh contrasts present in the opposing forces. The hatred between the parties[35] was such that it induced the Sienese not to take prisoners: the result was *il grande scempio che fece l'Arbia colorato in rosso*, "the great loss that stained the Arbia red", writes the White Guelph Dante (*Inf.* X, 85–86).

The consequences of the defeat could have been dramatic for Florence. The winners asked for the city to be razed to the ground, as had happened in Carthage after the Third Punic War, in 146 BC; but this did not happen due to the firm opposition of Farinata degli Uberti, who thus earned the admiration of Dante. In fact, while the poet placed him in *Hell* because of an alleged heresy, in reality he made Farinata a Homeric hero, consigning him to history.[36] For the next six years, from 1260 to 1266, Florence was governed by the Ghibellines who implemented an authoritarian policy with respect to the openings made by the Guelphs. However, fearing the revolt of the People, they bent to accept the mediation of pope Clement IV.

---

[35] A chronicler of the time placed an invocation to the Holy Mother made before the battle by one of the Sienese leaders with these words: *"Virgin Mary, help us to our great need, and free us from the hands of these lions and these superb men, who want to devour us"*.

[36] Dante describes the meeting of Forinata with these words: *"Io avea già il mio viso nel suo fitto; / ed el s'ergea col petto e con la fronte / com' avesse l'inferno a gran dispitto"* (I had already fixed my eyes in his; and he was rising up with his breast and forehead as if he had Hell in great disdain; *Inf.* X, 34–36).

He openly favored the Guelph party to the point that, in order to isolate Manfred, he excommunicated all the Ghibellines of Tuscany. It was a heavy move for a society based on trade, which made it possible to isolate the die-hards since it authorized the debtors to default. The emperor's paladins were then on the ropes and in 1266, when Manfred was defeated and killed in Benevento, the People rose up expelling the true Ghibellines out of Florence and appointing Charles of Anjou as *podestà* for a period of six years. It was in reality a surrender to a new master, of which many citizens were disappointed.

While the Count of Provence was taking possession of the south and center of Italy, effectively encircling the pope who had hoped to use him to his advantage, a period of great instability had begun in Germany, with the simultaneous presence of many pretenders to the imperial throne and the occult direction of the Vatican. Finally, in 1273 the German prince Rudolf I of Habsburg was designated as the new king of the Romans. It was a preliminary step to the imperial coronation. Rudolf was a character appreciated by the pontiff who counted on his lack of interest in the regions of Northern Italy and at the same time on the possibility of using his military weight to counterbalance the power acquired by Charles of Anjou in the same regions. But the crisis of Angevin hegemony caused by the revolt of the Vespers in Sicily in 1282, and the death of Charles himself three years later, brought back to the fore the Ghibellines. From Tuscany, Romagna, and the Duchy of Spoleto, they gathered in Arezzo and exiled the Guelphs from the city. The latter asked for protection from the Florentines, who were consolidating their hegemony in Tuscany.

In Florence, left without her most noble *podestà*, the People raised the head and new institutions appeared, designed for an expansion of the political participation. In 1282 the priory of Arts had been founded and 11 years later the city had assumed the official title of Republic, with new magistracies such as the *gonfaloniere di giustizia* with the functions of president (to use modern terminology) and the priors of the Arts, with executive power. The ancient ally and now rival, the Pisan Republic, had been put out of competition by a heavy defeat suffered in 1284 in the naval war with Genoa. And on June 11, 1289 in Campaldino, in the Upper Casentino, owing to the audacity of Corso Donati, the Florentine Guelphs definitively defeated the Ghibellines headed by Arezzo. The twenty-three-year-old Dante Alighieri also took part in the battle as a *feditor*, that is a lightly armed knight. Although a skilled horseman and expert in falconry, the poet did not distinguish himself for courage. Later he would write that he had "*a lot of fear, and in the end very great joy for the various cases of that battle*".

Florence was now firmly Guelph, but the traditional divisions, linked to social class and to wealth, remained, stressed further by the social reform made by Giano della Bella.[37] These divisions led to the creation of two warring factions. On the one side there were the Whites, led by the rich family of Cerchi, merchants enriched enormously and expression of a moderate pro-papal policy, who managed to govern from 1300 to 1301; on the other side the Blacks,[38] headed by the Donati family, representing an arrogant[39] financial and commercial aristocracy most closely linked to the interests of the Church. In 1300 pope Boniface VIII, the symbol of every evil for Dante,[40] sent to Florence his legate, cardinal Matteo d'Acquasparta, with the official task of restoring peace. But in fact, he had to downsize the power acquired by the Whites who were establishing alliances with the Ghibelline cities of Pisa and Arezzo. When the situation became critical, the municipality decided to send the leaders of the factions into exile. The provision, for which Dante was also responsible, displeased everyone: the Whites, who were in government, complained that they were included in the punishment, while the Blacks even attempted a coup. And when the exiled Whites were recalled to town, this partisanship gave Acquaspata the right to launch the anathema[41] on Florence. Then Boniface sent a new peacemaker, Charles of Valois, who took over the city for the faction of the Blacks. Among the Whites forced into exile there was Dante Alighieri.

2.13 To the end of our story

Before passing to Dante's life, it seem appropriate to have a quick look at the evolution of the Holy Roman Empire on the death of Frederick II, as for the poet this is the key institution to rule the Western world, and Italy in particular, in continuity with the glorious history of ancient Rome. According to the divine design, the Church has instead the role of moral

---

[37]Belonging to one of the oldest Ghibelline noble families, Della Bella had become a Guelph for political reasons. He was the champion of the most popular classes of Florence. In 1293 he issued a provision (*Ordinamenti di Giustizia*) which excluded the magnates from the government and gave power to the lower classes.

[38]The colors were inherited by those of the two parties into which the city of Pistoia was divided.

[39]One of the things that most irritated Corso, head of the Donati family, an arrogant and violent man, was the pomp with which the Cerchi lived in their luxurious Florentine mansion.

[40]While never quoting him directly, Dante places Boniface in *Hell*.

[41]Measure stronger than excommunication; the latter separates only from Christian communion, while the anathema separates from the Church itself.

guide that should never be corrupted by the thirst for earthly power. Dante blames the Church for her competition with the empire but also blames the emperor (Albert) for his inertia in the Italian politics:

> *Ahi gente che dovresti esser devota,*
> *e lasciar seder Cesare in la sella,*
> *se bene intendi ciò che Dio ti nota,*
>   *come esta fiera è fatta fella*
> *per non esser corretta da li sproni,*
> *poi che ponesti mano a la predella.*
>   *O Alberto tedesco ch'abbandoni*
> *costei ch'è fatta indomita e selvaggia,*
> *e dovresti inforcar li suoi arcioni,*
>   *giusto giudicio da le stelle caggia*
> *sovra 'l tuo sangue, e sia novo e aperto,*
> *tal che 'l tuo successor temenza n'aggia!*
>   *Ch'avete tu e 'l tuo padre sofferto,*
> *per cupidigia di costà distretti,*
> *che 'l giardin de lo 'mperio sia diserto.*
> (*Pur.* VI, 91–105).

"Ah, people who should be devoted and permit Caesar to sit in the saddle, if you attend to God's words to you, see how this beast has become savage, not being governed by the spurs, ever since you seized the reins. O German Albert, who abandon her, so that she becomes untamed and wild, while you should mount between her saddle-bows, may just judgement fall from the stars onto your blood, and let it be strange and public, so that your successor may fear it! For you and your father, held fast by your greed for things up there, have suffered the garden of the empire to be laid waste."

Albert of Habsburgs, mentioned by Dante, held weakly the royal title of *rex Romanorum* from 1298 to 1308 when he was assassinated. Before him and since the death of Conrad IV, son of Frederick II, in 1254, in Germany there was a sort of interregnum during which no king achieved universal recognition. With the loosening of the imperial yoke, the various German princes consolidated their independence from the central power. Similarly did the Italian communes. In 1273, Rudolf I of Germany had been elected, first king of the Habsburgs family, supporter of the Staufen house, then Adolf, count of Nassau. None of them could boast the title of emperor.

Finally, in 1312 Henry VII of Luxembourg was crowned emperor of the Holy Roman Empire in Rome after the failed attempt of Philip IV of France to seize the title of king of the Romans for his brother, Charles of Valois, with the help of Clement V, first French pope, who in 1309 had settled the apostolic see in Avignon. In spite of a very short reign, Henry had his own *Romzug* during which he reclaimed control of Italy, gaining with his actions the appreciation of White Guelphs like Dante and the historian Dino Compagni. But he failed to bend Florence. On his death in 1313, he left the Tuscan city even more aware of his power, king Robert of Naples arbiter of Italian politics, and in the north of the peninsula, two powerful imperial vicars, Cangrande I della Scala in Verona and Matteo Visconti in Milan.

Although still weak militarily (in the second decade of the XIV century it suffered two defeats in the open field by the Ghibellines), and although struck by a serious economic crisis due to the failures of families too exposed with their money loans, while Dante was approaching the end of his life, his city was then taking off as a regional power.

Chapter 3

Dante's biography and works

3.1 Birth and infancy

Dante was born in Florence in 1265. The exact date of birth is not know. However he says in the *Commedia* that he "first breathed the Tuscan air" (*Par.* XXII, 117) in the zodiac sign of Gemini, so between the 21st of May and the 21st of June. But, for the calendars at the times of Dante, the Gemini sign was beginning on the 18th of May. Of the mother we know only the name, Bella, and that she died soon, when Dante was a child. His father Alighiero (of Bellincione of Alighiero) died around 1283. He belonged to a family of small nobility, and was a business man, lending money. His great-great grandfather, Cacciaguida, had been knighted by Conrad III and died at the Second Crusade in 1147. Dante's name is a shortened version of Durante.

In canto XXV of the *Paradiso* Dante refers to Florence as the *"bello ovile ov'io dormi' agnello"*, "lovely sheepfold where I slept as lamb" (v. 5), and a little later he mentions the *"fonte del mio battesmo"*, "font of my baptism" (vv. 8–9), the baptistery of San Giovanni, a beautiful octagonal building standing in both Piazza del Duomo and Piazza San Giovanni, across from the Cathedral of Saint Mary of the Flower and the Giotto bell tower. The house of the Alighieri family was in the parish of San Martino del Vescovo in the district of Porta San Piero. In the same district lived the Portinari family. Dante mentions in the *Vita Nova* that at the age of nine he fell in love with Beatrice (Bice) Portinari who was one year younger: *"D'allora innanzi dico che Amore segnoreggiò la mia anima"*, "From then on I say that Love dominated my soul". However Beatrice married Simone de' Bardi around 1280 and died young in June 1290.

3.2   Youth

Dante had as teacher Brunetto Latini (about 1220–1294 or 1295), Florentine writer, poet, politician, and notary, who after the victory of the Ghibellines at Montaperti was exiled in France. There he wrote *Li livres dou tresor*, an encyclopedic work in Oïl, where he deals with universal history, medicine, physics, astronomy, geography, and architecture. The young Alighieri went also to Bologna to perfect his studies (he was there in 1287). As Giovanni Villani (1280–1348) wrote, he became *"rettorico perfetto tanto in dittare, versificare, come in aringa parlare"*, a perfect rethorician in dictating and in versifying, as well as in haranguing.

On June 11, 1289 Dante participated in the battle of Campaldino with the Florentine Guelphs against the Ghibellines of Arezzo. He was among the *feditores*, knights of the first rank, lightly armed, and remembers the battle in *canto* V of the *Purgatorio*, while talking to Buonconte da Monte-feltro, who had died there:

> *E io a lui: "Qual forza o qual ventura*
> *ti travï ò sì fuor di Campaldino,*
> *che non si seppe mai tua sepoltura?"*
> (*Pur.* V, 91–93)

"And I to him: 'What force or chance carried you away from Campaldino, so that your burial place was never known?'"

Dante started writing love poems in vernacular when he was eighteen, as an exchange with young friends. Among them were Guido Cavalcanti (1258–1300), Manetto Portinari (a brother of Beatrice) and Forese Donati (about 1250–1296, a cousin of his wife Gemma). These poems were collected in the *Vita Nuova* in 1292–1293 and much later, after Dante's death, in the *Rime*.

The marriage between Dante and Gemma Donati, member of an important Florentine family, was arranged in 1277 and actually happened around 1285. They had four children: Jacopo (about 1287–1348), Pietro (about 1300–1364), Antonia (?–1371; she became sister Beatrice) and Giovanni (around 1293–1353).

3.3   Public life

In order to participate in public life, in 1295 Dante enrolled in the art of doctors and apothecaries (*medici e spaziali*) and held various public offices between 1295 and 1302. From May to September 1296 he sat in the most important council of the city, the one of the Hundred; between June 15 and August 15, 1300 he was among the priors of Florence, elected precisely with the task of opposing the interference of pope Boniface VIII in the public life of Florence. Probably on his advice, the provision was taken to banish the leaders of the Whites and Blacks from Florence, following an assault by the Blacks against the consuls of the Arts and the reaction of the Whites. Among the banished was his friend Guido Cavalcanti and Corso Donati, his brother-in-law and facinorous and violent leader of the Blacks. While Dante was prior of Florence, he saved a child who was drowning in a baptismal font of the Baptistery of San Giovanni, where he had fallen. He remembers the event in canto XIX of the *Inferno*:

> *Non mi parean men ampi né maggiori*
> *che que' che son nel mio bel San Giovanni,*
> *fatti per loco d'i battezzatori;*
>   *l'un de li quali, ancor non è molt'anni,*
> *rupp'io per un che dentro v'annegava:*
> *e questo sia suggel ch'ogn'omo sganni.*
>   (*Inf.* XIX, 16–21)

"They seemed no less ample, nor greater, than those in my lovely San Giovanni, made as places for the baptizers; one of which, not many years ago, I broke because of one drowning inside it: and this be a seal to undeceive all men."

In November 1301, when Charles of Valois, the false peacemaker, approached Florence, Dante was in a delegation with two other ambassadors sent to Boniface VIII. The pope, having sent the two back to Florence, with the hope that they would intercede in favor of his thesis, kept the dangerous one, Dante. He was therefore not in Florence when Charles of Valois took the city and installed Cante de' Gabrielli from Gubbio as *Podestà* of the city, and when Cante, on January 17, 1302, first condemned the poet for corruption to pay a fine of 5000 libbre, in addition to the restitution of the extorted sums and to the confinement outside Tuscany for two years *pro bono pacis*, if he had payed. Then, since Dante had not come back nor payed, on March 10, 1302, he was sentenced to death at the stake (*igne comburatur sic quod moriatur*).

3.4 Exile

Thus begins Dante's exile, a capital event not only for his practical biography; in fact, he suffered deeply from it, and it clearly conditioned the further development of his mindset and his poetry. At first Dante fought together with the Whites and with them he hoped to be able to return to

Florence; but by July 1304, disgusted by the *compagnia malvagia e scempia* "evil and foolish company", he withdrew and made a party by himself. The *dolorosa povertade* "painful poverty" forced him to take advantage of the liberality of the various princes, thus merging with people of all sorts; he became, more or less, a "court man", painfully worried and disdainful of this condition. He hoped perhaps at first to be recalled to Florence, and to such purpose provided with letters and other writings to exonerate himself of the Ghibelline mark that was affixed to him, originating from the fact that the exiled Ghibellines joining the Whites. Dante also strived to raise his reputation as scholar, producing doctrinal works of wider commitment than love poetry.

During the years 1304–1307 he worked on the *De vulgari eloquentia* and the *Convivio*. Perhaps the idea of the *Commedia* started in 1307, either when he was taking up the poem ex-novo, or when he took up ancient designs and maybe old sketches. Then Dante's political thinking took its final form. He searched for the reason for Italy's problems and found it in discords; he saw the cause for these in the lack of one single civil power, that is, of the empire, for the emperors were distracted by the affairs in Germany, as well as in the usurpation of their powers by the Church. Dante, in short, from a moderate Guelph became almost a Ghibelline. The stages of the exile are not all known to us: the first refuge was with the Scaligeri in Verona; in 1306 he was with the Malaspina in Sarzana in upper Tuscany; he may have spent some time in Paris, although this is not confirmed.

When in 1310 Henry VII descended in Italy, Dante's hopes were rekindled and stayed high for three years. Finally the emperor was coming to Italy, well resolved to put an end to discord and to affirm his supreme authority as *rex pacificus*. Even the pope seemed well disposed: the papacy–empire dualism seemed finished. Dante wrote an epistle to kings, princes, and peoples of Italy: a shout of celebrations; then he ran to pay homage to the emperor who was wearing the iron crown in Milan (1311). But Florence, the Guelph Florence, resisted. Indeed the city was at the head of the Italian resistance. In March the poet wrote from the Casentino an epistle against the *scelleratissimi* "infamous" Florentines; in April, another epistle to Henry himself, so that he would not linger in the north of Italy. Meanwhile, he was excluded from the amnesty of 1311 (the so-called reform of Baldo d'Aguglione).

When Henry attacked Florence, the poet, however, did not take the weapons against the homeland; he stayed in Casentino, perhaps by the count of Battifolle. But the death of the emperor (1313) cut off all hope: if not immediately after, certainly very soon Dante returned to Verona, this time a guest of Cangrande. Maybe to the time of the major opposition to the emperor by Florence, by Robert of Anjou, and by pope Clement V, dates the work in which he more directly and orderly exposes his political ideal, his generous if utopian dream of a universal monarchy: the *Monarchia*.

Dante wrote another epistle, religious-political, in 1314 after the death of Clement V: it was directed to the cardinals. They were eloquently exhorted to help Rome, deprived of its two Suns, the pope and the emperor, so that they would at least elect an Italian pope, who would bring the pontifical seat back to the Eternal City from Avignon where it had been forcibly transferred in 1309. Dante's last letter was written in 1315, traditionally known as directed "to the Florentine friend": friends and relatives had exhorted him to return to Florence as he could, taking advantage of an amnesty, as long as he would accept certain humiliating formalities. Dante refused.

After the defeat of Montecatini (1315), Florence turned into confinement the death sentences of the least dangerous of the banished. This time too Dante refused to accept: so that on November 6 he was again sentenced to death, and this time together with his sons, who had then passed 14 years of age. We do not know for sure where Dante spent his last years, during which he had a certain tranquility of life, and that he used to finish his main poem. At least since 1318 he was in Ravenna, of which was lord Guido Novello da Polenta, Francesca's[1] nephew: perhaps he held there the chair of poetry and rhetoric.

A trip to Verona would be documented by the *Quaestio de aqua et terra* (Question about water and land), that Dante should have presented in that city in 1320; but some scholars have expressed doubts about Dante's authorship of this work. In the years of the stay in Ravenna dates the poetic correspondence with Giovanni del Virgilio, who exhorted Dante to write in Latin instead of vernacular and invited him to Bologna. Dante replied with two *Egloghe*, declining the invitation. There are doubts also on authenticity of these compositions, which represent the first attempt at bucolic poetry, on the threshold of humanism.

---

[1]Francesca da Polenta (1259–1285), married to Gianciotto Malatesta, fell in love with his brother Paolo. The two lovers were discovered and killed by Gianciotto. This story is told by Dante in canto V of the *Inferno*.

Dante died in Ravenna on September 14, 1321, probably of malaria, returning from an embassy in Venice. He was buried in a marble urn in the church of San Pier Maggiore, which later was renamed basilica of San Francesco. The Florentines repeatedly requested in vain the ashes of their poet, and had to content themselves with a cenotaph. The Ravenna tomb, restored in 1483, was covered by a small temple in the 18th century; the whole area around it was settled in 1936. Dante's family continued in Verona in the descendants of Pietro Alighieri. It merged in the 16th century with that of the Serego, who then took the double surname.

Chapter 4

Astronomy before Dante

4.1 An ancient science

Astronomy is the oldest of the natural sciences. Observations of the sky and recordings of either the main continuous phenomena, such as the succession of seasons, or transients, as for instance the eclipses or the passage of comets, began some tens of thousands of years ago, in the Upper Paleolithic, motivated by curiosity, by the human need for measuring time and orienting, as well as by superstition. The most ancient cultures identified in fact the celestial bodies with divinities and spirits in a kind of naive pantheism. Thus astrology and mythological cosmogonies were born. The first astronomers were shamans and priests and such remained even in the great civilizations of the Fertile Crescent in Mesopotamia and in the Egypt of the pharaohs. Later, more articulated forms of representation and interpretation of phenomena developed, which corresponded to new and organic conceptions of the world based on ways of asking questions and organizing answers according to a strictly philosophical disposition.

Traditionally the birth of philosophy is placed in the Greek colonies of the west coast of Anatolia, more precisely in the thriving city of Miletus. There in 585 BC Thales, placed by Dante among the *"great spirits"*,[1] demonstrated the predictability of celestial phenomena by anticipating the

---

[1]   *Poi ch'innalzai un poco più le ciglia,*
       vidi 'l maestro di color che sanno
       seder tra filosofica famiglia.
           Tutti lo miran, tutti onor li fanno:
       quivi vid'io Socrate e Platone,
       che 'nnanzi a li altri più presso li stanno;
           Democrito, che 'l mondo a caso pone,
       Diogenés, Anassagora e Tale,

occurrence of a total Solar eclipse: an event considered so mysterious and terrifying as to induce immediate peace to the armies of the Lydians and the Medes, caught by a sudden darkness during the fight. The phenomenon of the eclipse is used in the *Paradiso* both in canto II (which describes the ascent of Dante and Beatrice to the sky of the Moon), to refute a theory on the Moon spots,[2] and in canto XXIX, where an exceptional case of eclipse is presented, when at the Savior's death the night star miraculously retraced its steps to obscure the Sun (*Par.* XXIX, 97–99):

> *Un dice che la luna si ritorse*
> *ne la passion di Cristo e s'interpuose,*
> *per che 'l lume del sol giù non si porse;*

"One says the Moon turned back during Christ's passion and interposed itself so that the light of the Sun could not reach the Earth."

Here is another manifestation of the divine ability to alter the course of the wandering stars, already revealed when Jehovah stopped the Sun and the Moon to allow Joshua to complete the extermination of his enemies (*Joshua*, 10, 12–13).

A few decades after Thales, Anaximander, also a native of Miletus, besides dealing with tracing the known lands and measuring sky and time (he invented the gnomon to detect the height of the Sun above the horizon), theorized a first mechanical model of the world. According to the Presocratic philosopher, the Earth, in equilibrium in space without any support because it was equidistant from everything else, should have a cylindrical shape. Humanity lived and prospered on one of its circular faces, with a diameter equal to three times the thickness of the cylinder. With this ingenious invention, two opposing requirements were met in one fell swoop: the apparent flatness of the Earth's surface and its roundness suggested by the analogy with the profiles of the Sun and the Moon. These two celestial bodies, then, were imagined as holes opened on as many hollow cylindrical rings, filled with fire. Their rotation around the Earth somehow reproduced

---

> *Empedoclès, Eraclito e Zenone;*
> (*Inf.* IV, 130–138)

"When I lifted my brow a little higher, I saw the master of those who know, sitting among a philosophical company. All gaze at him, all do him honor: there I saw Socrates and Plato, standing closer to him, in front of the others, Democritus, who assigns the world to chance, Diogenes, Anaxagoras, and Thales, Empedocles, Heraclitus, and Zeno."

[2]See Section 9.5.

the appearances, including the lunar phases and the eclipses, interpreted as total or partial occlusions of the relative openings.

In his cosmogony, Anaximander introduced an infinite or unlimited or indefinite creative entity, the *apeiron*, which together with the (pre)concepts of dynamics of opposites, symmetry, and centrality of the Earth, would have represented as many paradigms for later thinkers of the Athenian school.

4.2 The dawn of heliocentrism: the Pythagoreans

Dante recognizes as the first *"amatore di sapienza"*[3] (lover of wisdom) the philosopher and mathematician Pythagoras, born on the island of Samos overlooking the coast of Anatolia, but moved around 530 BC to Magna Graecia, where he founded his philosophical school. Cities such as Crotone, Taranto, and Metaponto saw the birth and then the dispersion of an intriguing esoteric doctrine that placed the number, or rather the harmonic relationship between numbers and sounds of the musical scale, as the principle of being of reality. Pythagoreanism therefore developed a conception of the cosmos as number and harmony, within a mystical vision in which science was attributed the function of an instrument for the purification of the soul.

The search for harmony would then become the dominant reason for the thinking of many philosophers and scientists to this day. *"Philosophy [nature] is written in that great book which ever is before our eyes — I mean*

---

[3] I say then that long ago in Italy, around the beginning of the foundation of Rome, [...] there lived a very noble philosopher by the name of Pythagoras. [...] Before him those who sought knowledge were not called philosophers but wise men, as were the seven sages of antiquity, whose fame is still renowned, the first of whom was called Solon, the second Chilon, the third Periander, the fourth Cleobulus, the fifth Lindius, the sixth Bias, and the seventh Prieneus. When Pythagoras was asked whether he considered himself a wise man, he refused to accept the appellation for himself and said that he was not a wise man but a lover of wisdom. So it came to pass after this that everyone dedicated to wisdom was called a "lover of wisdom," that is, a "philosopher," for *philos* in Greek means the same as "love" in Latin, and so we say *philos* for lover and *sophos* for wisdom, from which we can perceive that these two words make up the name of "philosopher," meaning "lover of wisdom," which, we might note, is not a term of arrogance but of humility. From this word was derived the name of the act proper to it, "philosophy," just as from "friend" was derived the name of the act proper to it, namely "friendship." Thus we may see, considering the meaning of the first and second words, that philosophy is nothing but "friendship for wisdom" or "for knowledge"; consequently in a certain sense everyone can be called a "philosopher," according to the natural love which engenders in everyone the desire to know. (*Convivio* III, XI, 1–6)

the universe — but we cannot understand it if we do not first learn the lan-guage and grasp the symbols in which it is written. The book is written in mathematical language, and the symbols are triangles, circles and other geo-metrical figures, without whose help it is impossible to comprehend a single word of it; without which one wanders in vain through a dark labyrinth", would have written Galileo Galilei in 1623 (*Il Saggiatore*, VI); and in the mid-twentieth century, the British physicist Paul Dirac, one of the fathers of quantum mechanics: *"God used beautiful mathematics in creating the world"*.

Philolaus of Crotone was a follower of Pythagoras. He was responsible for the first systematization of the teacher's doctrines. Like Socrates, he lived in the V century BC. He built a cosmological system based on the cult of numbers and yet adhering to the evidences gathered with observa-tions, more complex than Anaximander's rather naive one. At the center of the universe there was no longer the Earth, but an eternal fire: a con-jecture motivated by mythical and religious reasons, but also by the need to account with greater fidelity for the main phenomena observed in the sky. Identified with the One, the origin of all numbers, and with the seat of the highest divinity (Zeus), the central fire was the cornerstone of the uni-form circular motions of ten bodies. First of all the Earth, conceived as a sphere both for ideological reasons (the sphere was considered the geomet-ric symbol of harmony) and to account for the change in the texture of the night sky with latitude. In order to reproduce phenomena such as the rising and setting of the Sun, Philolaus demanded that the orbit of the Earth be completed in 24 hours. Then attributing to it a rotation synchronous with the revolution, he also justified why it was not possible for men, perenni-ally relegated to the face opposite to the central fire, to enjoy its intense light.

A phantom anti-Earth circulated in an inner orbit, synchronous with the Earth and in a diametrically opposite position, so as to remain hidden from human sight. A non-existent body which, in Aristotle's opinion, Philolaus introduced just to bring the total to ten. The *tetraktýs* (a divine number for the Pythagoreans, who represented it as an equilateral triangle with four points on each side and one in the center, and on which they pronounced their ritual oaths) corresponded to the sum of the unity, of the first even number, of the first odd number (3, since for the Pythagoreans the One has in itself the nature of both even and odd) and of the first square, according to the mystical and esoteric attitude characteristic of Pythagoreanism. An attitude that has marked some strands of thought over the centuries, from

Christian numerology (from which Dante himself draws), to the Jewish Cabala and the Renaissance Neoplatonism.

Let us return to Philolaus' model. Proceeding outwards, beyond the anti-Earth there were the Moon and then the Sun, imagined as a large mirror capable of reflecting the light of the central source and orbiting the central fire in one year. The five planets (Mercury, Venus, Mars, Jupiter, Saturn) followed on circumferences with radii multiple of another divine number, 3, with periods consistent with observations. The whole system was surrounded by the firmament that was probably motionless. Outside this spherical world, another very pure fire was in contrast to the central one.

Although apparently bizarre, Philolaus' model possessed several characteristics of great modernity, including the denial of geocentrism (which does not however mean a heliocentrism in the manner of Aristarchus or Copernicus) and the search for numerical harmony, later recovered by Kepler and which, in the XX century, became *mutatis mutandis* a main instrument of theoretical physics. Think, for instance, of the aforementioned Nobel Prize in Physics, Paul Dirac, author of what is considered the most beautiful equation in physics. In 1963, almost on the verge of paradox, he wrote: "*This result is too beautiful to be false; it is more important to have beauty in one's equations than to have them fit experiment*" (*Scientific American* 208, 5).

4.3 Plato and Eudoxus: the preconceived model of the world

With Plato, the model of the world returned to being geocentric for purely ideological reasons. Born in 428 BC at the twilight of a period of great political, economic, and cultural splendor ensured by Pericles to the city, the Athenian philosopher, a disciple of Socrates, founded his cosmology on an original theory of ideas. In his opinion, these represented the true, eternal, and universal components of a metaphysical reality, the study of which was entrusted to dialectics.

Instead — he thought — material things constituted only appearances, that are imperfect portraits of ideas. Such an assumption justified why the phenomena of the sensible world could possibly deviate from ideal models, resting on the Pythagorean concept of an order based on number and harmony. Exposed in the *Timaeus*, perhaps the work of Plato that had the greatest influence on Medieval thought, the Platonic model contemplated a spherical Earth, motionless at the center of the world and made up of the four elements postulated by Empedocles of Acragas (earth, water, air, and fire). Both the Sun, reintegrated in its role of primary dispenser of light, and the firmament revolved uniformly around it, so as to reproduce the alternation of day and night and the annual motion of the star. The planets were also arranged in motion at constant speed on circular geocentric orbits. The whole system was wrapped by the sphere of the Fixed Stars, according to the orderly design of a divine architect, the Demiurge, who had created a compact cosmos putting order in the primordial chaos, giving it a soul and an eternal life:

> ... *Le cose tutte quante*
> *hanno ordine tra loro, e questo è forma*
> *che l'universo a Dio fa simigliante,*

"All things whatsoever have order among themselves, and this is a form that makes the universe resemble God," says Dante in *Par.* I, 103–105.

It is important to note the *a priori* choice of the circle for the orbits, of the sphere for their shapes, and of the uniformity for the motions, as symbols and consequences of perfection and eternity: preconceptions that after Plato would have remained hard to die. The purely cinematic model, with no pretense or stringent need to adhere to the evidence, failed to calculate the predicted positions of the wandering stars (hence the scarce reliability of the ephemerides[4]), especially when the planets exhibited a mysterious inversion of the ordinary direction of travel towards east relatively to the Fixed Stars. These retrogradations, which we now know are due to the parallactic effect of the orbital motion of the Earth (the apparent retrograde motion of the planet is explained by the displacement of the observation point during the Earth's revolution), could not find room in the Platonic order of motions.

A new and more evolved model of the planetary system was devised by Eudoxus of Cnidus. Physician, mathematician, and astronomer, born in a Doric colony of Anatolia at the end of the V century BC, he was a pupil of Plato but also of the Pythagoreans Archytas of Taranto and Philistion of Locri. Eudoxus conceived a geometric representation of the celestial world with the Earth immobile at the center and around it a system of rotating spheres with uniform angular velocity, some of which used for the transport of various celestial bodies and others, more numerous, with service functions. Through a careful choice of the inclinations of the different rotation

---

[4]Lists of positions on the sky of the moving stars, computed for different epochs.

axes, these spheres had to allow the happening of those motions that Plato's uniform circular motion alone did not even contemplate.

A total of twenty-six spheres arranged into functional groups (which would become thirty-three in a subsequent elaboration by Callippus of Cyzicus, who added a sphere for each of the three planets Mercury, Venus, and Mars, and two for the Sun and the Moon), enclosed by a last sphere assigned to the Fixed Stars: enough to reproduce, with the precision appropriate to the instruments then available, the paths of wandering stars. While this complex cosmic mechanism had a geometrical content higher than the previous ones, by keeping the distance of the planets constant from the Earth it failed dramatically in reproducing the variations of the apparent brightness of the planets.[5]

It would have been Aristotle, inspired by Eudoxus who had already partially broken with Plato's ideological paradigm while remaining tied to the mathematical approach of his Pythagorean masters, to transform a merely descriptive mathematical construction into a physical model.

4.4 Dante's astronomical sources

Before addressing the Aristotelian revolution, fundamental for Alighieri's work, it is worth mentioning some Greek philosophical schools that made indirect but profound contributions to cosmology, *"le tre sette della vita attiva, cioè li Epicurî, li Stoici e li Peripatetici,"* "The three schools of the active life: namely the Epicureans, the Stoics, and the Peripatetics" (*Convivio* IV, XXII, 15) and that the great poet met and became acquainted with mainly due to the mediation of Latin thinkers such as Cicero and Seneca (*"spiriti magni"*, "great spirits" of *Inf.* IV, 130–144, and *Convivio* III, XIV, 8).

Epicurus, born on the island of Samos in the IV century BC, was in the search for an inner peace free from supernatural conditioning. Founder of a materialistic doctrine, he believed that soul dies with the body (the Epicureans *"l'anima col corpo morta fanno"*, "make the soul die with the body", stated shortly Dante; *Inf.* X, 15). In the wake of the thought of

---

[5]This variation, which depends on the distance Δ of the planet from Earth, can be very significant. Let us consider, for instance, the case of Mars: at opposition, it is $\Delta_o = 0.5$ AU, while at conjunction $\Delta_c = 2.5$ AU. The square of the ratio Δ_c/Δ_o tells us that the brightness of Mars (nominally) changes by a factor of 25, that is by as much as 3.5 mag. As a general rule, the luminosity of a planet depends on its elongation (angular distance from the Sun) and it is maximum during retrograde motion, as this phenomenon occurs when the distance from Earth is the shortest.

Democritus and Leucippus, he proposed a conception of nature ruled by the casual meeting of indivisible particles, the atoms. Cantor of this philosophy, which was also a *modus vivendi* (a style of life), was the Latin poet Titus Lucretius Caro, who lived in the first century BC. Many scholars believe that Dante could not read it directly, if it is true that his monumental didactic work, *De rerum natura*, would have become widespread only starting from the year 1417, when it was accidentally rediscovered in a German convent by the humanist Poggio Bracciolini (see however Section 7.3). Probably only a few passages reported in his other readings reached Alighieri; but often these tastings were enough for his genious to get a personal and somehow complete idea on the subject (this is the case of the *Almagest*, as we shall see soon). After all, Immanuel Kant also conceived his *Universal natural history and theory of the heavens or essay on the constitution and the mechanical origin of the whole universe according to Newtonian principles* (1755) after reading a newspaper article on the model of the universe by the English astronomer Thomas Wright.

Inspired by the thought and figure of Zeno of Citium, who lived between the IV and III centuries BC, that of the Stoics was instead a philosophical school with a strong ethical, rational, and pantheistic imprint. On the cosmogonic level, it took its cue from the meditations of Heraclitus of Ephesus, raising fire to symbolize a constantly turning reality characterized by cosmic cycles of creation and destruction. A process of palingenesis governed by a universal reason, from which two principal sources arise: material things and the vital breath to animate them.

The peripatetics, so called because they discussed philosophy while walking under the arcades of the Lyceum (*perípatos* in Greek means walk), take us back to Aristotle of Stagira, teacher of Alexander the Great and *"maestro di color che sanno"*, "master of those who know" (*Inf.* IV, 131). His thought more than any other had an influence on Alighieri, even in the astronomical field. Dante had access to the original works of the Stagirite that survived time and contamination with Neoplatonic thought prevailing in the Upper Middle Ages. He also drew heavily on the analyses and compendia prepared by Arab scholars (and then translated into Latin in the towns that served as an interface between the Islamic and Christian worlds, such as Toledo and Córdoba) and on the revisions of the Dominican scholastic Thomas Aquinas and by his master, the German Catholic bishop Albert the Great and his school. Among the Arabs, Averroes, born in the Muslim Spain, stands out as the one who *" 'l gran comento feo"*, "made the great commentary" (*Inf.* IV, 144). He lived in the XII century and was so deeply conquered by Aristotle's immense knowledge to define him "*rule and [a] model that nature had conceived to show what the extreme perfection of man is*". Dante was in turn so fascinated by Averroes that he illegitimately placed him in Limbo reserved for the great pagans born before Christ saved humanity, thus blameless for not being Christians. For their merits they are allowed to live in a "*nobile castello*", "noble castle" full of symbols, "*sette volte cerchiato d'alte mura, / difeso intorno d'un bel fiumicello*", "seven times encircled by high walls, defended all around by a lovely little stream" (*Inf.* IV, 106–108), although far from God.

4.5 Aristotle, "*master of those who know*"

Concerning physics and astronomy, Aristotle broke with the previous tradition, and above all with "his friend" Plato,[6] preferring the "truth", that is the adherence to the appearances and at the same time the search for the causes, and, with deductive logic, the demonstration of the necessity of the effects. Statements such as the existence of two distinct worlds, each with its own chemistry and different physics (the corruptible and imperfect terrestrial world confined within the sphere of the Moon, the first of his complex cosmic clockwork, and the celestial one, perfect and eternally similar to itself), the absence of vacuum, the uniqueness of the cosmos, and the eternity of time, were all the results of preconceptions but also the consequences of a reasoning and of a confrontation with nature.

In *De coelo*, one of Aristotle's natural books, available at the time of Dante in the translation from Arabic into Latin made in the second half of the XII century by Gerard of Cremona, even the cosmological model of the homocentric spheres of Eudoxus loses its exquisitely geometric characteristic to acquire a physical connotation. The spheres, for example, from abstract geometric entities became real structures, albeit made up of ether, the fifth element absent on Earth and thus unprovable experimentally. The circularity and uniformity of the motions remained as fundamental principles. They ensured the achievement of perfection (understood as the absence of any change but cyclical) and therefore of

---

[6]The maxim "*Amicus Plato sed magis amica veritas*" is often attributed to Aristotle, as a paraphrase of the *Nicomachean Ethics*: "*Still perhaps it would appear desirable, and indeed it would seem to be obligatory, especially for a philosopher, to sacrifice even one's closest personal ties in defense of the truth. Both are dear to us, yet it is our duty to prefer the truth.*" Actually the concept was firstly coined by Plato himself (in the *Republic*) with reference to his master Socrates.

eternity (which is proper to the supreme entity), with an evident tres-
passing into a para-pantheism that could not but displease Christian
theologians.

It must be said that in the Greek world, with the notable exception
of the Stoics and of Plato who assigned an origin to the world, it was
widely believed that the Universe was immutable, uncreated, and eternal.
The rational motivation of this preconception (not different from the one
that would have led Albert Einstein to imagine a static universe and the
British cosmologist Fred Hoyle a stationary one, dynamically always equal
to itself owing to a mysterious spontaneous creation of matter from nothing)
lies in the difficulty of responding to the question: what was there before?
Thinkers were disturbed by the idea of a time before the time; a theme that
would have given rise to the profound reflections of Augustine of Hippo in
his *Confessions*,[7] up to those of the French philosopher Henri Bergson and
of the German Martin Heidegger.

4.6   The terrestrial world

We shall see how the cosmological system of Aristotle is constituted and how
it works, starting from the Earth. He modelled it as a sphere for reasons of
symmetry and in order to interpret some already known phenomena: the
change in the texture of the constellations with the latitude, a matching
with the images of the Sun and the Moon, and the shape of the shadow left
on during lunar eclipses. He believed, like the great geographers of his time,
that the known and inhabited lands constituted a sort of gigantic continent,

---

[7]I answer him who asks, "What was God doing before He made heaven and earth?"
I answer not, as a certain person is reported to have done facetiously (avoiding the
pressure of the question), "He was preparing hell," said he, "for those who pry into
mysteries." It is one thing to perceive, another to laugh — these things I do not answer.
For more willingly would I have answered, "I know not what I know not," than that
I should make him a laughing-stock who asks deep things, and gain praise as one who
answers false things. But I say that You, our God, are the Creator of every creature;
and if by the term "heaven and earth" every creature is understood, I boldly say, "That
before God made heaven and earth, He made not anything. For if He did, what did He
make unless the creature?" [...] I know that no creature was made before any creature
was made. (*Confessions*, XI, 14). For that very time You [Creator] made, nor could
times pass by before You made times. (*Confessions*, XIII, 15). What, then, is time? If
no one asks me, I know; if I wish to explain to him who asks, I know not. Yet I say
with confidence, that I know that if nothing passed away, there would not be past time;
and if nothing were coming, there would not be future time; and if nothing were, there
would not be present time. (*Confessions*, XIV, 17).

the *ecumene*, crossed by big rivers: according to the historian of the V
century BC, Herodotus of Halicarnassus, a sort of quadrangle grouping
Europe, Western Asia, and North Africa around the Mediterranean sea.
Herodotus's work had a fundamental influence in the ancient world. In the
II century, the Alexandrian Claudius Ptolemy developed this Earth model
by giving a unitary arrangement to the geographic knowledge; among other
things, he provided Dante with the basis for the magnificent construction
of the places of the Inferno and of the Purgatorio.

According to Aristotle, the order of the world as well as geocentrism
itself were accounted for by the theory where each one of the four elements
constituting the sublunary world moves, in absence of impediments, to
reach its natural place. The tendency of the heavy elements (earth and
water) is in fact to fall downwards; that of the light ones (fire and air) to
rise upwards. The four homocentric sublunary spheres are therefore, from
top to bottom: fire, air, water, and earth, according to a conception which
Dante also refers to (see, *e.g.*, the first part of canto XII of the *Inferno*).
It is a naive model of gravity, developed in Aristotle's *Physics* and part of
an original approach to motion, from which important consequences would
have derived. Some of them, blatantly wrong, had a negative influence on
the development of natural philosophy for the centuries to come. Let us
see in some detail how and why.

Adopting the thesis reported by Plato in the *Timaeus* and rejecting
that of the atomists, Aristotle believed that matter was constituted by the
dynamic combination of the four primary elements of Empedocles, each
characterized by a pair of the four fundamental qualities: dry/humid and
cold/hot. Each body acquires the characteristic of the dominant element
among those that compose it; for example, a wet stone is of earth-type.
This is a relevant fact since Aristotle distinguished two types of motions in
the sublunar world: natural and violent (forced). The former are straight
(even if not uniform, as is reasonable in a corruptible world) and they
express the tendency of the primary elements to reach their natural places.
We find an example in the *Commedia* when Beatrice explains to Dante
(*Par.* I, 115) that: "*Questi ne porta il foco inver' la luna*", "This carries
fire on up toward the moon". The allusion is to the instinctive ascent of
fire towards the lunar sky, to which the poet attributes a cause, the God
"*che tutto move*", "who moves all things" (v. 1). Regarding the second type
of motions of the terrestrial world, the Stagirite erroneously believed that
a force should be applied to a body in order to keep its velocity constant;

this mistake would have been corrected about two millennia later by Galileo Galilei.

The logical consequences of these preconceptions are numerous: firstly the justification of the centrality of the Earth; then, in controversy with the Democritean physics, the negation of vacuum (*horror vacui*), imposed by the need of the primary elements to always know the direction of their natural places (a sort of Mecca that does not admit uncertainties and duplications). This same reason made it inadmissible the existence of the infinite worlds predicated by Epicurus (recognizing that the existence of several Earths would imply giving conflicting messages to the elements in search of their natural places; see also Chapter 14).

4.7   The celestial world

As for the motions of the Earth, in addition to the revolution proposed by Philolaus, Aristotle consistently excluded also the diurnal rotation, because the velocities involved would have been too large to escape observation. Actually, he knew an approximate value of the Earth's radius; with which the rotational velocity of the Earth, if any, would have taken an abnormally large value for the time, without however leaving any apparent sign of this speedy motion. He therefore preferred to attribute the peculiar behavior to the firmament, in the awareness of the extraordinary nature of the celestial structure and the concomitant impossibility of carrying out an experimental verification.

The Earth, cradle of that humanity whom Aristotle would have dealt with by addressing issues such as ethics, aesthetics, logic, and politics, was the center of a system of seven main spheres, one for each of the wandering bodies. They were placed in a succession of distances determined by the value of the apparent motion: Moon, Mercury, Venus, Sun, Mars, Jupiter, and Saturn, plus a last sphere for the Fixed Stars. Almost a table of contents for the canticle of Dante's *Paradiso*, albeit the veneration of the poet for *"the Philosopher"* breaks down in this passage from the *Convivio* precisely on the number of celestial spheres:

> *Aristotile credette, seguitando solamente l'antica grossezza delli astrologi, che fossero pur otto cieli, delli quali lo estremo, e che contenesse tutto, fosse quello dove le stelle fisse sono, cioè la spera ottava; e che di fuori da esso non fosse altro alcuno. Ancora credette che lo cielo del Sole fosse immediato con quello della Luna, cioè secondo a noi. E questa sua sentenza così erronea può vedere chi vuole nel secondo Di Cielo e Mondo, ch'è nel secondo de'*

libri naturali. Veramente elli di ciò si scusa nel duodecimo della Metafisica, dove mostra bene sé avere seguito pur l'altrui sentenza là dove d'astrologia li convenne parlare.[8] (*Convivio* II, III, 3–4)

Actually these two theses (the eight spheres and the contiguity of the sky of the Sun to that of the Moon) have no parallel in Aristotle's work, but are found in his commentators, including Albert the Great, one probable source for Dante.

The Aristotelian spheres are not geometric abstractions but, as we said, real physical structures. The first sphere of ether, the one that carries the Moon with it, also acts as a separator and border between the two worlds of the dichotomous Aristotelian cosmos. In the celestial world, time is not linear but cyclic (an indispensable presupposition of an eternal extraneousness to time and its failures, as befits the Platonic Demiurge) and the natural motions are circular and uniform. These strong hypotheses accounted for the apparent invariability and perfection of the skies.

For each of the wandering planets, the system of motions was similar to the purely geometric one of Eudoxus: not a single sphere, but groups of four (three for the Moon and the Sun), of which two to reproduce the diurnal and annual motions and a mirror pair to create the back and forth of retrograde motions, all conveniently oriented in order to optimize adherence with observations (*ad hoc* adjustments and corrections that would be considered unacceptable in a modern cosmological theory). These packages of spheres, which in Eudoxus' model functioned independently of the others, in the Aristotelian cosmology instead had to find a physical connection so that each could draw its own movement from a single engine; a sort of support point for the lever that "*move il sole e l'altre stelle*", "moves the sun and the other stars" (*Par.* XXXIII, 145). For this reason Aristotle, after updating Eudoxus' model by eliminating some superfluous spheres, introduced numerous others, for a total of 55, twenty-two more than those

---

[8] Aristotle, merely following the longstanding ignorance of the astrologers, believed that there were only eight heavens, of which the outermost, containing the whole, was the one on which the stars are fixed, namely, the eighth sphere, and that beyond it there was no other. Moreover he believed that the heaven of the Sun was contiguous to that of the Moon, that is to say, was second from us. Anyone who wishes can find this extremely erroneous opinion of his in the second book of his *Heaven and the World*, which is in the second of the books about nature. However, he excuses himself for this in the twelfth book of the *metaphysics*, where he plainly shows that he was only following the opinion of others where he was obliged to speak of astrology. (From *Digital Dante*, https:// digitaldante.columbia.edu/text/library/the-convivio/)

of Callippus. The aim was to establish the physical connections between the various groups, while maintaining for each one the independence of the motions from the others. This was possible by making sure that each movement was offset by an opposite movement, as in a kind of cosmic Penelope's shroud.

The resulting construction was complex and still unsatisfactory in terms of the variations in brightness exhibited by the planets over time, necessarily attributable to variations (see note at page 59) in distance which, however, the model did not contemplate in order not to shatter the dogma of the perfection of the skies. The apparently simpler element was represented by that single sphere of the Fixed Stars which enclosed the cosmos within a finite volume. It had the sole tasks of reproducing the diurnal rotation (subtracted from the immobile Earth) and of communicating motion to all other spheres at the urging of an Unmoved Mover. The sphere of the Fixed Stars, in fact, did not constitute the ultimate boundary, beyond which for Aristotle — as we have said — neither empty space nor other worlds could exist. Outside there was a divine intelligence, a first cause or prime mover which, in a completely mechanical way and without any direct intervention, communicated to the spheres, conceived as animated, a sort of desire to move. In the words of Beatrice (*Par.* XXVII, 106–111):

> La natura del mondo, che quïeta
> il mezzo e tutto l'altro intorno move,
> quinci comincia come da sua meta;
>   e questo cielo non ha altro dove
> che la mente divina, in che s'accende
> l'amor che 'l volge e la virtù ch'ei piove.

"The nature of the world, which stills the center and moves all the rest around it, here begins as from its goal, and this heaven has no other *where* than the mind of God, in which is kindled the love that turns it and the power that it rains down."

The progress of observation technologies, with astrometric and photometric estimates gradually more accurate (although made by eye[9]) and

---

[9]The human eye is a very precise instrument in measuring differential luminosity, but very poor for absolute estimates.

referring to an ever wider time base, highlighted the need of reviewing the dogma of the pure circularity of motions.

The paradigm shift occurred with Hipparchus, one of the greatest and most complete astronomers of the Greek school. Born at the start of the II century BC in Nicaea, a city in Anatolia (*Asia Minor*) that later became famous for the Council of 325 AD, so fraught with consequences for astronomy too, he was the author of extraordinary discoveries and decisive methodological innovations. Perhaps his best known contribution is the finding, through the comparison between his measurements of the positions of some stars and those of the same objects made much earlier by the Babylonians, that the celestial pole is not fixed. It in fact participates in a conical motion resulting in the so-called precession of the equinoxes (a discovery with implications for astrometry and also for the calendar).

Hipparchus dealt with geography, where he introduced a methodology for measuring longitudes based on lunar eclipses, coming to imagine — the geographer Strabo tells us — the existence of an unknown continent in the middle of the Ocean (the enormous river encircling the world). He is credited for the development of trigonometry, for an accurate method of predicting the eclipses of Sun and Moon, and for having devised the astrolabe, an instrument later popularized by the Arabs. He built a refined star catalog by introducing the famous six classes of magnitudes that would have been the archetype of astronomical photometry up to present days. He observed the sudden appearance of a nova star in Scorpio, which led him to reflect on the actual immobility of the skies. He also measured the distance of the Earth from the Moon, with a result, however, less accurate than it had been obtained a century earlier by another astronomer native of Samos, Aristarchus.

Around the middle of the third century BC, Aristarchus had stunned the Athenians by proposing two ingenious methods for determining the distances of the Earth from the Moon and from the Sun. For the first, he used the transit time of our satellite through the shadow of the Earth during an eclipse; a phenomenon whose nature had been understood already by the Milesian philosophers. The result depended on the knowledge of the radius of the Earth, which Eratosthenes of Cyrene, rector (we would say today) of the Mouseion of Alexandria in Egypt, would soon after ensure with amazing precision by exploiting the different inclination of the solar rays at the summer solstice in the two localities of Alexandria and Siene (now Aswan). A measure that was not only the result of a brilliant idea, but of a set of notions and hypotheses on the shape of the Earth and on the behavior of light. As for the distance of the Sun, Aristarchus proposed a conceptually correct but practically impossible experiment, which led to a far too short estimate.

Rightly famous for these two results, he is also known, together with Heraclides Ponticus (IV century BC), to have been the most convinced supporter of a heliocentric hypothesis which, by placing the Sun at the center of the world, assigned to the Earth both a diurnal rotation and an annual revolution. But the time was not ripe for a solution that would have had great difficulties affirming itself even eighteen centuries later, with Copernicus, as it was blamed of driving man out of the center of the world. Aristarchus was not believed and perhaps, following a legend related by the Italian poet Giacomo Leopardi, he was even exiled on charges of impiety (a true infamy, we would comment today condemning any retaliation against opinion crimes, but much better than ending at the stake as it happened to Giordano Bruno on February 17, 1600).

Let us return to Hipparchus. The Nicaean astronomer sought a way out of the problem of eliminating the constancy of the distances of wandering stars while preserving the dogma of uniform circular motions. He found the solution in a mechanism devised by Apollonius of Perga, active between the III and the beginning of the II century BC: a distinguished mathematician at the Library of Alexandria and true tightrope walker of those plane curves called conics (ellipse, parabola, and hyperbola) because they are obtained by cutting a double cone with a plane.

Hipparchus' idea is the following. Imagine a uniform circular motion of a point P on an epicycle of radius r around a center that in turn rotates uniformly on a deferent of radius R and center O (see Fig. 4.1). It is easy to believe that, with this construction, the distance between P and O varies remaining between $(R - r)$ and $(R + r)$. Egg of Columbus! By placing the planets on epicyclical motions, Hipparchus had saved the Pythagorean–Platonic preconceptions as well as the appearances (at the price — William of Ockham[10] would have said — of a further multiplication of entities; but, while cumbersome, a similar approach would have been used at the beginning of the XX century to model the spiral pattern of disk galaxies). Concepts that Dante also used when writing about the sky of Venus (the third heaven) and carnal love (see Section 10.2).

---

[10]English Franciscan friar, scholastic philosopher, and theologian, author of the homonym principle: *Entia non sunt multiplicanda sine necessitate* (Entities are not to be multiplied without necessity).

4.8 Claudius Ptolemy and mathematical syntax: a model of the world that lasted fourteen centuries

Over six centuries of cosmological speculations found their synthesis in the work of a mysterious scientist who lived in Alexandria in the second Hellenistic period, that following the Roman invasion of Egypt: Claudius Ptolemy. He was an astronomer, astrologer, geographer, and the author of the *Mathematical Syntax*, one of the most influential scientific works of antiquity. This book was a reference point for Dante who, however, did not read it in the original but likely in the Latin version translated from Arabic, as can be deduced from the quotations he makes of it in the *Convivio*. Known by the name subsequently assigned to it by the Arabs, the *Almagest* (the Greatest) collects, reorganizes, and integrates all the previous mathematical and cosmological knowledge, in particular the models of the world of Aristotle and Hipparchus. The result is an almost perfect instrument in its performance and even not too complicated if compared with the convoluted heliocentric machinery devised by Copernicus 14 centuries later, who however started from the right hypothesis.

The main assumptions of Ptolemaic cosmology, enshrined in the *Mathematical Syntax*, are that: (1) the sky moves like a sphere; (2) the Earth, according to our perception, has a spherical shape in all its parts, and (3) is at the center of the celestial sphere and in relation to it may be treated as a point. The most notable innovations concern the introduction of corrective actions to epicyclic motion, such as the equant (Fig. 4.1) which, with exquisite but unjustified elegance, added the necessary (albeit cyclic) variability to a combination of uniform motions. Specifically, Ptolemy claimed that the centers O of the planetary deferents did not coincide with the points from which the motions appear uniform and that the Earth, from which these motions are observed, was located in a direction diametrically opposite to O and at an equal distance from the center. The equant (an attempt to model what Kepler would succeed to do with his second law), would have been questioned by Arab astronomers and abandoned in the Middle Ages.

It is interesting to note how the periods of the epicyclic motions of the planets outside the sphere of the Sun (Mars, Jupiter, and Saturn) and those of the centers of the epicycles along the deferents of the inner planets (Mercury and Venus) coincided with the apparent annual motion of the Sun. It would have been enough to reflect on that to reach the Copernican

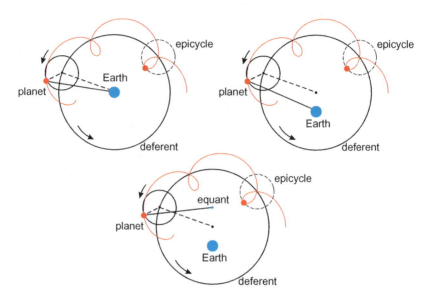

Fig. 4.1 Sketch of the planetary models based on (*upper-left*) epicycle and deferent, (*upper-right*) eccentric, and (*lower-center*) equant. Courtesy of E.Y. Bannikova and M. Capaccioli.

solution fourteen centuries in advance, sacrificing however the reassuring anthropocentric ideology.

Ptolemy also introduced in his model an integration to the Aristotelian immobile mover, a sphere called the crystalline sky, to account for the precession of the equinoxes discovered by Hipparchus. In Dante's words:

> *Tolomeo poi, acorgendosi che l'ottava spera si movea per più movimenti, veggendo lo cerchio suo partire dallo diritto cerchio, che volge tutto da oriente in occidente, constretto dalli principii di filosofia, che di necessitade vuole uno primo mobile semplicissimo, puose un altro cielo essere fuori dello Stellato, lo quale facesse questa revoluzione da oriente in occidente: la quale dico che si compie quasi in ventiquattro ore, [cioè in ventitré ore] e quattordici parti delle quindici d'un'altra, grossamente asegnando.*[11]
> (*Convivio* II, III, 5; cf. *Vita Nova*, 19, 5).

---

[11]Later Ptolemy, perceiving that the eighth sphere moved with several movements (since he saw that its circle deviated from the true circle which turns everything from east to west) and constrained by the principles of philosophy, which necessitated the simplest primum mobile, supposed that another heaven existed beyond that of the Fixed Stars which made this revolution from east to west, a revolution that, I say, is completed in about twenty-four hours (that is, in twenty-three hours and fourteen out of fifteen parts

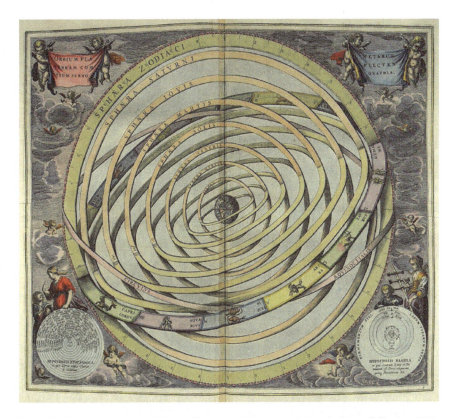

Fig. 4.2 The geocentric world from *Harmonia Macrocosmica* by Andreas Cellarius, 1661. (Reproduced from Wikipedia)

The Arab and Christian philosophers added a further extension to the Aristotelian heaven, the Empire, to host angels, blessed (the *"candida rosa"* "white rose" of the *"milizia santa"* "holy militia", *Par.* XXXI, 1–2) and God himself. They did not do much else on a theoretical level, other than abolishing the equant and improving the parameters of the model with refined observations. However, they produced wonderful treatises and compendia, which constituted precious sources for Dante: think of the Islamic philosophers Al-Battani, al-Farghani, Avicenna, Averroes, and Alpetragius. They also made original contributions to mathematics (by inventing

---

of another, roughly speaking). (From *Digital Dante*, https://digitaldante.columbia.edu/text/library/the-convivio/)

algebra, which is a word derived from the Arabic *al-gabr*, meaning reduction, and by introducing the concept of zero, developed by the Indians), physics, optics, and, with al-Biruni, to geography.

4.9   Can we say that Dante was a scientist?

Before closing this section, let us ask ourselves what Dante knew about all this matter, remembering that he was not a scientist in the modern sense or even in a broad sense, but mainly a great medieval poet and a man of genius in search of God. As a scholar, he had formally access to the literature of his time, albeit with considerable limitations, considering the fact that he was forced to spend most of his life wandering from place to place at a time when the libraries, even those of the powers he frequented, were largely incomplete and deficient (metal movable-type printing press would have been introduced in Europe by Johannes Gutenberg only 130 years after Dante's death). From the point of view of geography and cosmology, his reference was Ptolemy, although, as we have already noted, he had not read the works of the scientist from Alexandria in the original and probably only knew them through the Latin translation of a reduction made by an Arab scholar (a consideration — we want to stress it again — that does not diminish, but rather exalts the greatness of a man capable of grasping the general sense of an idea from some fragments).

Dante certainly read Thomas Aquinas and Albert the Great, whom he placed in Paradiso (*Par.* X, 97–99):

> *Questi che m'è a destra più vicino,*
> *frate e maestro fummi, ed esso Alberto*
> *è di Cologna, e io Thomas d'Aquino.*

"He who is closest to me on my right was brother and master to me, and he is Albert of Cologne, and I Thomas of Aquino."

He also read Augustine of Hippo and his disciple Orosius, the Italian theologian and bishop Peter Lombard, "*quel Pietro fu che con la poverella/offerse a Santa Chiesa suo tesoro*" "that Peter who with the poor widow offered up his treasure to Holy Church" (*Par.* X, 107–108), and above all Aristotle, the most cited among the ancient scholars. Dante recovered his figure of a scientist even in comparison to the giant Ptolemy, despite the criticism of the Parisian school thinkers to some Aristotelian propositions deemed limiting the divine power. He collected and reworked this knowledge in the *Convivio*, an encyclopaedic compilation in "Italian" (the "volgare", an amalgamated literary language mostly derived from the dialect spoken in Tuscany), structured in treatises commenting on songs written by Dante himself; a project that he then abandoned to embark on the composition of the *Comedy*.

The physicist Paul Dirac, making fun of the passion for Dante of another physicist, Robert Oppenheimer, father of the atomic bombs dropped on Japan in 1945, said to him: *"How can you deal with physics and poetry at the same time? In physics we try to explain something unknown in simple terms. In poetry it is the exact opposite"*. Faced with these words, one would think that Dirac had never read the *Comedy*. If he had done so, he would have realized that Alighieri's poetry is the harmonious garment for a scientific treatise with no equal, the sum of knowledge and fruitful intuitions.

Chapter 5

The calendar of the Comedy

5.1 The date and the duration of Dante's journey

Dante was certainly not a person willing to leave something to chance. Even the departure of his journey to the Universe, however imaginary, had to take place at a particular moment. On this point all commentators agree. Instead, they disagree on setting the date, also because Dante did not tell it explicitly but, as he often did, left many clues; we are especially interested in the astronomical ones. The most important are these: it had to be a full moon, Venus was visible in the morning in the constellation of Pisces, the Sun was in the sign of Aries, and Mars and Saturn in the constellation of Leo.

Most commentators place the date in the spring of 1300, the first jubilee year in history, between mid-March and April 8, the day of the full moon. However, this choice does not satisfy some important astronomical conditions, especially, as we will see, the position of Venus. Therefore others have proposed that the departure of Dante's journey was a year later, in the night between March 24 and 25, 1301, which is compatible with all the positions of the stars. Furthermore, in canto XXI of *Inferno* the devil Malacoda says that 1266 years have passed since the death of Christ (vv. 112–114). Dante in the *Convivio* wrote that the Messiah died in his thirty-fourth year, therefore in the year 35, because he was born in the year 1, since the year 0 does not exist. Therefore: 1266 + 35 = 1301. March 25 was also believed to be the date of Christ's death.

With this solution, however, the coincidence with the jubilee year would be lost, which should have represented a particularly important symbol for the poet. The way out lies in the fact that, due to the calendar in force in Florence at the time of Dante, the year began on March 25, so the night between March 24 and 25, 1301 was still in the jubilee year announced by Boniface VIII. We, as astronomers, obviously prefer the date of March 24–25, 1301 and use the time scale marked by the Gregorian calendar.

Dante was lost in the forest on the night between March 24 and 25, 1301 and spent the whole of the following day there, and in the "*selva oscura*" he met Virgil. They entered together Hell on the evening of March 25. The descent into the funnel of Hell to reach the center of the Earth lasted only one day, until the evening of March 26 (Palm Sunday in 1301); the times refer to Jerusalem, located in the center of the emerged lands, therefore on the fundamental meridian of the time.

On the other side of the center of the Earth, the times are instead referred to the mountain of Purgatory, which was at the antipodes of Jerusalem; therefore, as for a change of time zone, there is a jump of twelve hours and one passes instantly from the evening of March 26 to the morning of the same day. The ascent that ended at the base of the mountain of Purgatory lasted about twenty hours, until just before dawn on March 27. Then Dante reached the top of Purgatory in just over three days, at noon on March 30.

The journey through Heaven probably lasted a day and a half, until midnight on March 31, 1301, which was Good Friday. All in all, from entering Hell to the Empyrean, the journey ended in less than seven days: not bad for someone who had to cross the whole Universe! Indeed, from the center of the Earth, which is also the center of the Universe, to its outer border, Dante took less than six days, from the morning of March 26 to midnight of 31.

5.2   The error of the Julian calendar

The recording of the passage of time, of the alternation of light with dark, and of the succession of the seasons, is crucial for mankind and concerns both the so-called linear time, advancing at a steady rate, without hesitations, from past to future (for example, the age of each of us, which flows continuously between the two extremes of birth and death), and the circular one, which is characteristic of the motions of celestial bodies (both true and apparent) and of the hands of the clock. It is from the circular time that we derive notions such as those of day and year, apparently simple but in fact complicated by the vagaries of the celestial motions to which they refer. For example, in the meaning that is most familiar and useful for everyday life, the day represents the time interval between two successive culminations of the Sun at noon, when the star reaches its maximum height on the horizon. We call it precisely "solar day", to distinguish it from another similar time interval.

The "sidereal day", also generated by the rotation of the Earth around its axis, considers, instead of the Sun, the culminations of the stars or, as astronomers say it, two consecutive meridian transits of that particular point of the sky where the two maximum circles of the celestial equator (perpendicular to the Earth's rotation axis) and of the ecliptic (plane of the apparent annual motion of the Sun) meet and where the Sun is in its ascending phase. Called the γ point or Aries point (if you notice it, the third letter of the Greek alphabet looks like a goat's head), it cannot be seen directly with the eyes since there is no fixed object in it; however it is very important in the measure of time.

If we divide the solar day into 24 parts, called hours, and use this unit for the sidereal day, we notice that this latter lasts approximately 23 hours and 56 minutes. Why is there such a difference? The reason is that, while it rotates on itself, the Earth also completes a revolution around the Sun in about 365 days (360 for the Babylonian astronomers who for this reason divided the rotation angle into 360 parts, called degrees, so that one degree was equal to the arc of a circle swept by the star in one day). It is easy to verify that $1/365$ of a day results in just the 4 minutes that separate the shorter sidereal day from the solar day.

Dante was fully aware of the consequences of the phenomenon, but obviously he had an erroneous idea of the causes. His astronomical culture came from reading the comments to Ptolemy's *Almagest* and contemplated a fixed Earth at the center of the universe, surrounded by the Sun, by the planets, and by the stars, all moved by complicated combinations of circular and uniform motions. An incorrect point of view, therefore. But, with marginal differences, the result was the same as that of the Copernican heliocentric model, which would have come to light more than two centuries after the poet's death.

The situation is similar with regard to the year. At the time of Dante, the Julian calendar was used; it had been promulgated by Julius Caesar in 46 BC, and had introduced a leap year of 366 days every four years, thus assuming that the duration of the average solar year was 365 days and a quarter. But, what's it about?

Let's start by saying that the solar (or tropic) year frames the complete cycle of the seasons and therefore marks a primary rhythm of nature, fraught with consequences for the existence of living beings on Earth. It is equal to the time taken by the Sun to return to the same solstice during its ascending–descending motion along the local meridian (again it is different from the sidereal year, which concerns an apparent net rotation of the Sun of 360°, due to the phenomenon of precession of the equinoxes).

Hipparchus of Nicaea was the first in the West to precisely measure the duration of the solar year, in the II century BC. According to what was transmitted to us by Ptolemy and by Albatenius, the aforementioned Arab scholar of the IX century, whose work was translated into Latin in the first half of the XII century by Plato di Tivoli (also called Plato Tiburtino) and therefore known to Alighieri, the astronomer of Nicaea established that the period between two solstices lasted "365.25 days minus 1/300 of a day-and-night", or 365.246 days: a value extraordinarily close to today's 365.2422 (there is a difference of just 5 minutes), and in any case very different from the round figure of 365 days adopted in the cultured Egypt of the pharaohs and of the renowned Library of Alexandria.

It is perfectly understandable that it is convenient to make the year consist of a whole number of days, but the cut of the decimals with respect to the actual duration introduces a systematic and consistent delay of the calendar with respect to the seasons (equal to about one day every four years). This is a problem of a practical but also religious nature, given the links between heaven and divinity. In order to solve it, the astronomer-priests of Ra provided occasional corrections, similar to those made on a not-too-precise analog clock.

To eliminate this arbitrariness, the astronomer Sosigene, Cleopatra's favorite, suggested to Caesar the simple workaround of adding one day every four years, in order to automatically take into account of a significant part of the neglected decimals. The great warrior, who rested in the arms of the beautiful queen, was familiar with the problem of time for having been *pontifex maximum* in Rome, so he took charge of the reform by imposing it on the territories of his vast domains. A step forward, inherited by the Church of Christ as it replaced the imperial power; to the point that the adjustment of the fundamental date of the vernal equinox, a metaphor for rebirth and creation, took place in 325 AD in Nicaea on the occasion of the first ecumenical council ordered by Constantine the Great to smooth out the dogmatic differences unleashed by Arianism and to affirm its hegemonic role within the Church.

Actually, however, the tropical year is shorter by eleven minutes and fourteen seconds than the 365 days and a quarter established by the Julian calendar, which therefore anticipated by one day every 128 years, as already mentioned. At the time of Dante the difference had risen to ten days from 46 BC; the vernal equinox, which the Council of Nicaea in 325 AD had set for March 21, fell in fact seven days earlier, on March 14. The thing was also known to the great poet, who mentioned it in canto XXVII of *Paradiso* with a beautiful hyperbole by Beatrice, who, to indicate a very distant time in the future, refers to when the spring equinox, due to error of the Julian calendar, will have moved to December 31, thus "unwintering" all January: "*Ma prima che gennaio tutto si sverni / per la centesma ch'è là giù negletta*", "But before January altogether unwinters itself with the hundredth they neglect out there" (vv. 142–143), where the hundredth part is actually 1/128. For this to happen, the equinox must be eighty days earlier, so 10,240 years must pass since the Council of Nicaea.

The error of the Julian calendar also had an effect on the determination of the date of Easter, which is set for the first Sunday after the first full moon of spring. By mistaking the date of the beginning of spring, there was the risk of mistaking the date of Easter. For this reason, the Church had to deal with the question, almost two and a half centuries after Dante's death, and by virtue of new and more precise measurements of the length of the year carried out by Arab and Jewish astronomers (they compiled the famous Tables commissioned by Alfonso X of Castile and León, published at the beginning of the second half of the XIII century) and by Copernicus himself.

Pope Gregory XIII set up a commission of experts, composed of the German astronomer Christopher Clavius, the mathematician Ignazio Danti from Perugia, and the physician and astronomer Luigi Lilio from Calabria, who proposed to the pontiff to make secular years no longer leap years, except for multiples of 400, thus setting the duration of the solar year at 365.2425 days, with an error of only 26 seconds. In 1582 Gregory XIII promulgated the reform, which took his name, with the bull *Inter gravissimas*, meaning that the matter was part of the most important provisions of the Holy Father. To recover the advance accumulated by the Julian calendar, ten days were eliminated from October 5 to 14, 1582, while maintaining the weekly sequence intact. The Gregorian calendar is still in effect today. It will cause a delay by one day with respect to the seasons in the still distant 4905 AD.

5.3   The date of creation

Who better than Adam knows when the world was created, given that he, as the Bible tells us, was shaped by God at the end of the week of creation? Dante, who met Adam in canto XXVI of *Paradiso*, asked him about it and the first man replied:

> *Quindi onde mosse tua donna Virgilio,*
> *quattromila trecento e due volumi*
> *di sol desiderai questo concilio;*
> * e vidi lui tornare a tutt'i lumi*
> *de la sua strada novecento trenta*
> *fiate, mentre ch'io in terra fu'mi.*
> (*Par.* XXVI, 118–123)

"Down there whence your lady sent Virgil, for four thousand, three hundred and two turnings of the sun I yearned for this assembly, and I saw him return along the road of all his lights nine hundred thirty times, while I lived on Earth."

For the creation Adam indicated first the 4302 years of his waiting in Limbo, that is, from his death to the descent of Christ into hell, and then the 930 years of his life (those also mentioned in *Genesis*). To these we must add the 1266 years that passed from the death of the Nazarene on the cross to Dante's journey. In total, 6498 years have passed. Very few compared to the current estimate of the age of the Universe (13.7 billion years), but in line with the story of the Bible.

Chapter 6

The Earth of Dante

Geographical knowledge at the time of Dante was limited, both because it did not include the Americas or southern Africa, and because in the east it stopped at the Ganges. Furthermore, while the latitude of the places was known with sufficient precision, as it is linked to the altitude of the North Star above the horizon, which was easy to measure, the same could not be said for longitude. In fact, its estimate required an accurate measurement of great distances on land, which was very difficult, and of time on the seas, for which there were no suitable instruments (they would arrive only four centuries later, thanks to the English inventor and watchmaker John Harrison).

Dante's Earth is spherical and has a radius of 3250 miles (*Convivio* **II**, VI, 10). Considering that the Florentine mile corresponded to about 2000 meters, the size of the Earth known by Dante was quite accurate, given that the actual radius is 6371 kilometers. The emerged lands were concentrated in the northern hemisphere and centered on Jerusalem; from there they extended in longitude 90° west to Cadiz, 90° east to the mouth of the Ganges, and in latitude 30° south to equatorial Africa and about 30° north to northern Europe (Fig. 6.1). Dante said in canto XXVII of *Paradiso* (vv. 79–84) that the extension in longitude of the Mediterranean, from Cadiz to Jerusalem, was 90°, when he noted that the Sun was at the meridian of Jerusalem and at the same time at the horizon in Cadiz.

Already at the beginning of canto XXVII of *Purgatory*, the poet had noticed that, at the same time as the Sun rose in Jerusalem, it was midnight in Spain, noon at the Ganges and the star was setting in Purgatory, confirming that these four places were located at 90° longitude from each other:

Fig. 6.1 Map of the emerged lands, as they were known in Dante's time.

Sì come quando i primi raggi vibra
là dove il suo fattor lo sangue sparse,
cadendo Ibero sotto l'alta Libra,
* e l'onde in Gange da nona rïarse,*
sì stava il sole; onde 'l giorno sen giva,
come l'angel di Dio lieto ci apparse.
    (*Pur.* XXVII, 1–6)

"As when he shoots the first rays to where his maker bled, when Ebro falls beneath high Libra and the waves of Ganges are scorched by noon, so stood the sun; and thus day was fading, when the glad angel of God appeared to us."

The differences in longitude used by Dante are overestimated by a factor of about two: the extension in longitude of the Mediterranean from Cadiz to Jerusalem is not 90°, but only 42°; the one from Jerusalem to the mouth of the Ganges is 55°. Since, as already noted, the size of the Earth was well known in the XIII century, this overestimate of the differences in longitude implies an overestimate of the distances by a factor of about two. This is quite surprising if you think that the length of the Mediterranean must have been known at the time with a good approximation. Perhaps Dante did not consider this problem and in any case he privileged the centrality of Jerusalem and the symmetry of the emerged lands.

6.1 The infernal abyss

Dante described the Inferno as an immense conical chasm with the axis coincident with the vertical of Jerusalem. It was produced at the beginning of time by the ruinous fall of Lucifer, the angel who had rebelled against God. Precipitating from the Heaven, this symbol of absolute evil remained stuck at the vertex of the cone, located at the center of the Earth (which is also the center of the world) (Fig. 6.2).

The entrance to the afterlife world is through a gate bearing the famous inscription:

> *Per me si va ne la città dolente,*
> *per me si va ne l'etterno dolore,*
> *per me si va tra la perduta gente.*
> *Giustizia mosse il mio alto fattore:*
> *fecemi la divina podestate,*
> *la somma sapienza e 'l primo amore.*
> *Dinanzi a me non fuor cose create*
> *se non etterne, e io etterno duro.*
> *Lasciate ogne speranza, voi ch'intrate.*
> *(Inf.* III, 1–9)

"Through me the way into the grieving city, through me the way into eternal sorrow, through me the way among the lost people. Justice moved my high maker: divine power made me, highest wisdom, and primal love. Before me were no things created except eternal ones, and I endure eternal. Abandon every hope, you who enter."

Inside the gate there is the *Antinferno* (Vestibule of Hell), a place reserved to the unclassified souls of those who in their lives did not take any party either for the good or for the bad. In order to reach the true Inferno, the lost souls need to cross the Acheron river with the boat of the ferryman Charon, "*un vecchio, bianco per antico pelo*" (*Inf.* III, 83). What follows are nine increasingly narrow circles. The first circle of the abyss, called Limbo, where Virgil himself resides, is reserved to unbaptized and virtuous pagans (and also some intruders, put there for no other reason than a particular affection of the poet). The other circles, with increasing severity of the punishment, are managed by Minos who tells the incoming

Fig. 6.2 Michelangelo Caetani, Overview of the Divine Comedy, 1855. (Reproduced from Wikipedia)

sinners where to go using his long tail:

> *Stavvi Minòs orribilmente, e ringhia:*
> *essamina le colpe ne l'intrata;*
> *giudica e manda secondo ch'avvinghia.*
> *Dico che quando l'anima mal nata*
> *li vien dinanzi, tutta si confessa;*
> *e quel conoscitor de le peccata*
> *vede qual loco d'inferno è da essa;*
> *cignesi con la coda tante volte*
> *quantunque gradi vuol che giù sia messa.*
> *Sempre dinanzi a lui ne stanno molte;*
> *vanno a vicenda ciascuna al giudizio;*
> *dicono e odono, e poi son giù volte.*
>   (*Inf.* V, 4–15)

"There stands Minos bristling and snarling: he examines the soul's guilt at the entrance; he judges and passes sentence by how he wraps. I say that when the ill-born soul comes before him, it confesses all; and that connoisseur of sin sees which is its place in Hell; he girds himself with his tail as many times as the levels he wills the soul to be sent down. Always many stand before him; each goes in turn to judgement, they speak and hear and are cast into the deep."

At the bottom of the conical abyss there is the frozen lake of Cocytus where Satan is imprisoned, committed to crushing the traitors: Judas, Brutus, and Cassius (the two organizers of the conspiracy to assassinate Julius Caesar). This is a singular place, a "*punto / al qual si traggon d'ogne parte i pesi*" (the point towards which the weights all move from every direction, *Inf.* XXXIV, 110–111) said Dante, noting a property of gravity, the absence of weight in the geometric center of the spherical Earth. This concept would later be taken up by Leonardo da Vinci in a famous thought experiment described in the *Codex Forster II* and made a scientific notion in 1687 by Isaac Newton (see Section 6.4).

Also Galileo Galilei, who was born almost 250 years after Dante's death, was quite accustomed to the work of the poet and dealt with the structure of Hell. There still exists a very rare copy of the *Commedia* that belonged to the Pisan scientist, with the dedication by the Vallombrosan abbot Orazio Morandi, who gifted it to him in 1624. It is an edition printed in Venice in 1555, the first one with the adjective "Divine" in the title (Fig. 6.3).

Between the end of 1587 and the beginning of the following year, the 23 years old Galilei even held two consecutive lectures at the Florentine Academy about the figure, site, and size of Dante's Inferno. It must be

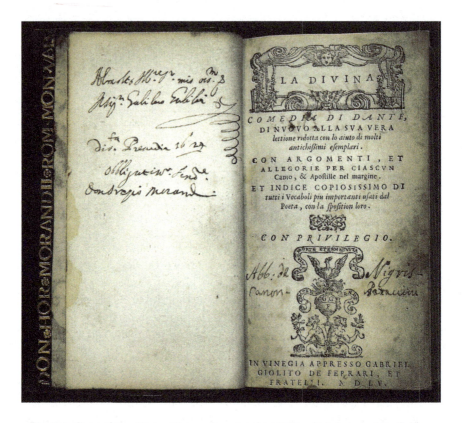

Fig. 6.3 Copy of the *Divina Commedia* printed in 1555, which belonged to Galileo.

said immediately that this was not a tribute to the great poet by a brilliant emerging intellectual but rather a performance of a skilled and ambitious young man in search of accommodation (which he finally found at the University of Padua in 1592, having failed in 1587 the appointment in Bologna). Galilei had been urged by academics to judge two conflicting models of Dante's Inferno.

The first one was by Antonio di Tuccio Manetti, a pupil of Brunelleschi (himself a great lover of Dante), a mathematician, architect, and leading figure in the Florence of his years. He had carried out a true geometric investigation of the infernal building, looking at the structure with the new criteria of perspective. His considerations were published in 1481 in the introduction to a commentary on the *Commedia* edited by the humanist Cristoforo Landino, tutor of the young Lorenzo the Magnificent, and adopted as true by the Florentine intellectuals.

The other reconstruction of the infernal measures was produced 63 years later, in 1544, by Alessandro Vellutello, a scholar from Lucca who had moved to Venice, also as part of a commentary on the *Commedia* (rich of beautiful engravings). He argued that Hell should have linear dimensions 10 times smaller than those adopted by Manetti. It was a classic academic problem, transcending the poetic power and grandeur of Dante's picture. However it had to be solved as a consequence of Vellutello's polemical attitude towards Florentine intellectuals, flattened on the description given by Manetti.

Galilei was therefore asked to act as arbiter and he did so on his part. He began with a first lesson in which, with the help of graphs of which we have no trace, he presented the characteristics of Manetti's model: shape, size, and location of the infernal chasm and then number, extent, and interval of the circles through which it developed, recognizing the fidelity to Dante's text. He then devoted the second lesson to underlining and refining the criticisms of Vellutello, such as the fragility of the covering of an excessively large infernal cone. Galilei falsified the objection by using for this purpose the very solid dome of the Duomo of Florence, whose dimensions he re-scaled to those of Inferno, thus performing a brilliant mathematical exercise but an illegal change of scale from the physical point of view (of which he would have made count later).

6.2 The position of the *selva oscura*

The *selva oscura*, the dark forest, where Dante got lost at the beginning of the *Commedia* and from where he entered the inverted cone of Hell, was located on the edge of the cone itself. According to the reconstruction made by Antonio Manetti and Galileo (see the previous section), the base diameter of the infernal cone would be equal to the terrestrial radius; so 3250 miles, as Dante said in the *Convivio*. Therefore the forest would be located about 1600 miles from Jerusalem, which was on the axis of the cone. It is interesting to note that this distance roughly corresponds to that estimated at the time between Jerusalem and Naples, at about twenty kilometers from Cumae, where Virgil had brought Aeneas into the Avernus with the Sibyl. Dante obviously imitated his teacher.

6.3 The *folle volo* of Ulysses

In order to undestand Dante's geography, it is imperative to refer to the story that Ulysses made of his last journey (the *folle volo*, the mad flight) into canto XXVI of *Inferno*:

> *Io e ' compagni eravam vecchi e tardi*
> *quando venimmo a quella foce stretta*
> *dov'Ercule segnò li suoi riguardi*
>   *acciò che l'uom più oltre non si metta;*
> *da la man destra mi lasciai Sibilia,*
> *da l'altra già m'avea lasciata Setta.*
> (*Inf.* XXVI, 106–111)

"I and my companions were old and slow when we came to that narrow strait which Hercules marked with his warnings so that one should not go further; on the right hand I had left Seville, on the other I had already left Ceuta."

The Homeric hero and his men, then old, arrived at the Pillars of Hercules, with Seville on the right and Ceuta on the left, and decided to continue westwards into the unknown ocean, turning left, then south:

> "*e volta nostra poppa nel mattino,*
> *de' remi facemmo ali al folle volo,*
> *sempre acquistando dal lato mancino.*
>   *Tutte le stelle già de l'altro polo*
> *vedea la notte, e 'l nostro tanto basso,*
> *che non surgëa fuor del marin suolo.*
>   *Cinque volte racceso e tante casso*
> *lo lume era di sotto da la luna,*
> *poi che 'ntrati eravam ne l'alto passo,*
>   *quando n'apparve una montagna, bruna*
> *per la distanza, e parvemi alta tanto*
> *quanto veduta non avëa alcuna.*
>   *Noi ci allegrammo, e tosto tornò in pianto;*
> *ché de la nova terra un turbo nacque*
> *e percosse del legno il primo canto.*
>   *Tre volte il fé girar con tutte l'acque;*
> *a la quarta levar la poppa in suso*
> *e la prora ire in giù, com'altrui piacque,*
>   *infin che 'l mar fu sovra noi richiuso*".
> (*Inf.* XXVI, 124–142)

"and, turning our stern toward the morning, of our oars we made for the mad flight, always gaining on the left side. Already all the stars of the other pole I saw at night, and our own pole so low that it did not raise above the floor of the sea. Five times renewed, and as many diminished, had been the light beneath the Moon, since we had entered the deep pass, when there appeared to us a mountain, dark in the distance, and it seamed to me higher than any I had seen. We rejoiced, but it quickly turned to weeping; for from the new landa whirlwind was born and struck the forequarter of the ship. Three times it made the ship to turn about with all the waters, at the fourth to rise its stern aloft and the prow to go down, as it pleased another, until the sea had closed over us."

Ulysses and his mates passed the equator; in fact, the pole star disappeared below the horizon and they saw the stars of the southern pole instead. After five months of travel, when they sighted the mountain of Purgatory, a storm swept over their ship and swamped them.

When Dante arrived by another way to the shore of Purgatory, he would remember that those places had never been seen by anyone who then returned to report them, obviously thinking back to Ulysses and implicitly to his own fortune as an imaginary explorer who could tell his adventure:

> *Venimmo poi in sul lito diserto,*
> *che mai non vide navicar sue acque*
> *omo, che di tornar sia poscia esperto.*
> (*Pur.* I, 130–132)

"Then we came on to the deserted shore, which never saw any man sail its waters who afterwards experienced return."

To indicate the five months of his last journey, Ulysses said that the "light beneath the Moon" had turned on and off five times, referring to the thin crescent moon that faces downwards near the new moon, and is even more so in the equatorial areas he had crossed.

The poet placed Ulysses in the *bolgia* of the fraudulent, however acknowledging him for "*l'ardore* / [...] *a divenir del mondo esperto* / *e de li vizi umani e del valore*", "the ardor [...] to gain experience of the world and of human vices and worth" (*Inf.* XXVI, 97–99) and above all putting in his mouth the wonderful exhortation to his companions: "*fatti non foste a viver come bruti,* / *ma per seguir virtute e canoscenza*", "you were not made to live like brutes, but to follow virtue and knowledge" (*Inf.* XXVI, 119–120). Perhaps Dante, in facing his journey through the Universe, felt close to Ulysses in the ardor of knowledge, and in giving the dead their place in the afterlife he shared the experience of human vices and virtues. However, there remains a substantial difference between Ulysses and Dante: if the sea closed on the pagan, preventing him from continuing the journey to Purgatory, the "predestined" Christian was allowed to reach the top of the mountain and to climb unharmed to the Empyreon.

6.4 Crossing the Earth's center

At the bottom of Hell, the center of the Earth, which was also the center of the Universe, Lucifer was wedged. Dante and Virgil descended along his body:

> Quando noi fummo là dove la coscia
> si volge, a punto in sul grosso de l'anche,
> lo duca, con fatica e con angoscia,
>   volse la testa ov'elli avea le zanche,
> e aggrappossi al pel com'om che sale,
> sì che 'n inferno i' credea tornar anche.
> (*Inf.* XXXIV, 76–81)

"When we came to where the thigh is hinged, exactly at the widest of the hips, my leader, with labor and difficulty, turned his head to where he had his shanks, and clung to the pelt like one who climbs, so that I supposed we were returning into Hell again."

At the passage of Lucifer's hips, the force of gravity suddenly reversed and what had been a descent turned into an ascent. This can be explained in the framework of Aristotelian physics, which incorporates gravity in the theory of natural places, for which, as already mentioned, heavy bodies tend by nature to move towards the Earth, which is the center of the Universe. Isaac Newton, four centuries after Dante, would describe gravity as an attractive force of all bodies that have mass, and then Einstein, another two centuries later, would interpret it as a mass-induced deformation of space–time.

6.5 The island of Purgatory

Dante said that Purgatory was the highest mountain on Earth and that its top was at the upper limit of the clouds:

> Per che non pioggia, non grando, non neve,
> non rugiada, non brina più sù cade
> che la scaletta di tre gradi breve;
>   nuvole spesse non paion né rade,
> né coruscar, né figlia di Taumante,
> che di là cangia sovente contrade;
>   secco vapor non surge più avante
> ch'al sommo d'i tre gradi ch'io parlai,

dov'ha 'l vicario di Pietro le piante.
(*Pur.* XXI, 46–54)

"For this reason no rain, no hail, no snow, no dew, no frost falls any higher than the little stairway with but three steps; clouds do not appear, whether dense or thin, nor lightning, nor the daughter of Thaumas [Iris, the personification of the rainbow], who often changes neighborhoods down there; dry vapor does not rise above the highest of the three steps I mentioned, where Peter's vicar sets his feet."

In *Naturalis historia* (II, 85) Pliny reported that Posidonius had set to forty stages the limit at which the clouds and winds reached; this is equivalent to 7400 meters. A calculation of the height of Purgatory can also be made from the time elapsed between sunset at the base of the mountain and that at the top, which obviously occurs later due to the curvature of the Earth's surface. From a passage in *Purgatorio* (XXVII, 1–6, 67–69) it can be estimated that this time difference is about a quarter of an hour, so the height of the mountain is about 10,000 meters, quite in agreement with the previous estimate.

Dante did not specify the dimensions of the base of the island, but these can be derived from the height and the average inclination of the slope. In canto IV of the *Purgatorio* the poet said that near the base the slope was steeper than the climbs that lead to San Leo (a fortress in Romagna), to Noli (a fortress in Liguria), to Bismantova (a cliff in Emilia) and on Cacume (a mountain in Lazio), then steeper than 45°:

> *Vassi in Sanleo e discendesi in Noli,*
> *montasi su in Bismantova e 'n Cacume*
> *con esso i piè; ma qui convien ch'om voli;*
> (*Pur.* IV, 25–27)

"One can go up to San Leo and descend to Noli, one can climb to Bismantova or up the Cacume with one's feet, but here one must fly."

Higher up the slope becomes less steep: the inclination can be estimated at around 30° from the fact that the Sun's rays are parallel to the slope, as Dante suggested in canto VI of *Purgatorio*, and from the time of the day at that moment. Higher still the slope decreases further until the Earthly Paradise at the top, which is horizontal. So, if the average slope were 30°, a height of 10,000 meters would correspond to a base diameter of 35 kilometers and a circumference of 110 kilometers, roughly the same as the island of Gran Canaria, which however is only 1956 m high, one fifth of Purgatory.

6.6   The snow on the Apennines and the shadowless lands

In canto XXX of *Purgatorio*, after a reproach from Beatrice, Dante vented his pain by crying and assimilated his mood to the snow on the Apennines:

> Sì come neve tra le vive travi
> per lo dosso d'Italia si congela,
> soffiata e stretta da li venti schiavi,
>   poi, liquefatta, in sé stessa trapela,
> pur che la terra che perde ombra spiri,
> sí che par foco fonder la candela;
>   (*Pur.* XXX, 85–90)

"As snow on the living beams along the back of Italy turns to ice, driven and compressed by the Slavic winds, and then, liquefied, trickles into itself as soon as the land that loses shadow breathes, so that it seems fire melting the candle."

The snow on the Apennines freezes in winter, hit by the winds that come from the land of the Slavs, i.e. by the cold winds from the north-east (the *bora*); then it melts and pours on the underlying layers due to the southern winds, which come from the lands where the shadows are shorter, because the Sun is more vertical. Dante used here another simile full of geographical connections.

6.7   Buggea and Marseille at the same longitude

In canto IX of the *Paradiso* Dante showed off his geographical knowledge by making Folchetto say that Marseille, the city where the bishop-troubadour was born, is located at the same longitude as Buggea (Bougie in French or Béjaïa in Arabic) on the coast of Algeria. This is approximately correct; in fact, Marseille is only about 20′ to the east of Buggea. Since, as we have already mentioned, the accurate measurement of longitudes was not possible in the Middle Ages, Dante was probably referring to nautical charts of the time, which were relatively precise thanks to the experience of sailors.

Chapter 7

Atmospheric phenomena

The first sphere above the Earth is that of water; it is followed by the sphere of air, in which, according to Dante, the atmospheric phenomena that we deal with in this chapter occur. Some, such as lightning, shooting stars, comets, and even the Milky Way (which we will consider separately in Section 12.1) occur on the border with the next sphere of fire.

7.1 The rainbow

Dante spoke several times of the rainbow. For example, in canto XXIX of the *Purgatorio*, to describe the bright colors of the luminous trails left by the seven candelabra carried in procession, shortly before the meeting with Beatrice, he referred to the rainbow and to the moon halo (the Moon, mythologically assimilated to Diana, the Greek Artemis born in Delos, is called Delia):

> *e vidi le fiammelle andar davante,*
> *lasciando dietro a sé l'aere dipinto,*
> *e di tratti pennelli avean sembiante;*
> *sì che lì sopra rimanea distinto*
> *di sette liste, tutte in quei colori*
> *onde fa l'arco il Sole e Delia il cinto.*
> (*Pur.* XXIX, 73–78)

"and I saw the flames move on, leaving the air behind them painted, and they seemed like brushes drawn along, so that the air overhead was marked with seven stripes, all in those colors with which the Sun makes his bow and Delia her belt."

This metaphor is scientifically very appropriate, because in both cases the colors are due to refractions within drops of water. Dante in canto XXV

of the *Purgatorio* mentioned the fact that the rainbow is due, as we well
know, to the refractions of the Sun's rays in the raindrops:

> *E come l'aere, quand'è ben pïorno,*
> *per l'altrui raggio che 'n sé si reflette,*
> *di diversi color diventa addorno;*
> (*Pur.* XXV, 91–93)

"And just as the air, when it is very moist, becomes adorned with various
colors because it reflects another's rays."

Again in canto XII of the *Paradiso* the poet, to describe the double
crown of spirits that appeared during the speech of Thomas Aquinas, used
the double arc of the rainbow:

> *Come si volgon per tenera nube*
> *due archi paralleli e concolori,*
> *quando Iunone a sua ancella iube,*
> *nascendo di quel d'entro quel di fori,*
> *a guisa del parlar di quella vaga*
> *ch'amor consunse come sol vapori,*
> *e fanno qui la gente esser presaga,*
> *per lo patto che Dio con Noè puose,*
> *del mondo che già mai più non s'allaga:*
> (*Par.* XII, 10–18)

"As through a tenuous cloud two arcs curve parallel and colored alike, when
Juno commands her handmaid, the outer born from the inner one, like the
speech of that desirous nymph whom love consumed as the Sun does vapors,
and cause people here to predict the weather, thanks to the pact God made
with Noah, that the world will never again be flooded."

This representation of the phenomenon is very accurate. It occurs with
the Sun behind the observer, who instead has the rain in front, so the
sunlight is generally projected onto the clouds that produce the rain. In
addition, the two arcs are concentric, that is parallel, and the colors are
repeated in reverse ("the outer born from the inner one"), since the external
arc is due to a double reflection in the raindrops. The nymph Iris, who was
Juno's handmaid, was accompanied by the nymph Echo, who was consumed
with love for Narcissus. To these pagan memories Dante added the biblical
one of the rainbow, which marked God's reconciliation with humanity after
the punishment of the universal flood and sealed the pact with Noah that
there would be no other total floods.

7.2 The wind in the divine forest

At the top of the mountain of Purgatory Dante entered the Earthly Paradise and ventured into the divine forest of Eden, where a sweet wind blew:

> *Un'aura dolce, sanza mutamento*
> *avere in sé, mi feria per la fronte*
> *non di più colpo che soave vento;*
> *per cui le fronde, tremolando, pronte*
> *tutte quante piegavano a la parte*
> *u' la prim'ombra gitta il santo monte;*
> *non però dal loro esser dritto sparte*
> *tanto, che li augelletti per le cime*
> *lasciasser d'operare ogne lor arte;*
> (*Pur.* XXVIII, 7–15)

"A sweet breeze, unchanging in itself, struck my brow with non greater force than a gentle wind, by which the pliant branches, trembling, were bent, all of them, toward where the holy mountain casts its earlier shadow, but not parted so much for their straightness that the little birds in the treetops left off exerting their every art."

It was a light wind that blew in the direction in which the mountain of Purgatory casted its shadow in the early morning, then towards the west, and, while moving the branches of the trees, it did not disturb the birds. This wind that blows on top of Purgatory, therefore at an altitude of about 10,000 meters and at a latitude of $-31°$, with constant speed and at the antipodes of Jerusalem (see Chapter 6), brings to mind the jet stream that is at the same height and latitude. These are atmospheric streams that blow with constant speed and direction at latitudes of about $\pm30°$ (tropical jet stream) and $\pm60°$ (polar jet stream). But unlike the aura of the Earthly Paradise, they go from west to east and are much faster, around 200 km/h. They have a great influence on air navigation and for some routes they are used by pilots to save fuel.

7.3 The atmospheric dust

In canto XIV of the *Paradiso* Dante described the spirits of the heaven of Mars moving along a cross, comparing them to the particles of atmospheric dust that danced in a ray of light in the darkness:

> *così si veggion qui diritte e torte,*
> *veloci e tarde, rinovando vista,*
> *le minuzie d'i corpi, lunghe e corte,*

> *moversi per lo raggio onde si lista*
> *talvolta l'ombra che, per sua difesa,*
> *la gente con ingegno e arte acquista.*
> (*Par.* XIV, 112–117)

"thus down here we see moving, straight and oblique, swift and slow, always changing, some long, some short, the tiny motes in the sunbeam that sometimes stripe the shade that with wit and art people make for their defence."

In Dante's triplets, which masterfully render the phenomenon, a passage from Lucretius echoes:

> *Contemplator enim, cum solis lumina cumque*
> *inserti fundunt radii per opaca domorum:*
> *multa minuta modis multis per inane videbis*
> *corpora misceri radiorum lumine in ipso*
> *et velut aeterno certamine proelia pugnas*
> *edere turmatim certantia nec dare pausam,*
> *conciliis et discidiis exercita crebris.*
> (*De rerum natura* II, 114–120)

"In fact, observe every time the rays infiltrate / pouring the light of the Sun into the darkness of the house: / in the empty space you will see many minute bodies mix / in many ways in the same light of the rays / and as in eternal contest to assault, battle, / fighting in squadrons, and never stop,/ constantly pushed by unions and splits."

Lucretius's words "many minute bodies [...] in many ways" are found in Dante with "straight and oblique, swift and slow, [...] some long, some short, the tiny motes". As we have already noted, according to many scholars, the Florentine poet was not directly acquainted with *De rerum natura*, which was widespread only after the rediscovery by the humanist Poggio Bracciolini in 1417 AD. On the other hand, it can be thought that Dante had direct access to Lucretius's work because, as demonstrated by Giuseppe Billanovich, the Latin poem circulated in its entirety from the second half of the XIII century in the cenacle of the Paduan prehumanists that flourished around Lovato de' Lovati.

Another hypothesis is that Alighieri knew the text of Lucretius through Lactantius, a Latin writer who lived at the turn of the III and IV centuries, who in the *De ira Dei* (X) reported with comments on some passages from *De rerum natura*.

7.4 The shadow of the Earth at the horizon

To describe the light of the angels that slowly disappeared as he ascended to the Empyrean, Dante at the beginning of canto XXX of the *Paradiso* used as a simile the stars that gradually disappear with the progress of the aurora, when the Earth casts its shadow almost horizontally:

> *Forse semilia miglia di lontano*
> *ci ferve l'ora sesta, e questo mondo*
> *china già l'ombra quasi al letto piano,*
>   *quando 'l mezzo del cielo, a noi profondo,*
> *comincia a farsi tal, ch'alcuna stella*
> *perde il parere infino a questo fondo;*
>   *e come vien la chiarissima ancella*
> *del sol più oltre, così 'l ciel si chiude*
> *di vista in vista infino a la più bella.*
> (*Par.* XXX, 1–9)

"Perhaps six thousand miles away the sixth hour is burning, and our world is lowering its shadow down almost to the level bed, when the transparency of the sky, deep above us, begins to become such that some stars no longer appear as far as this floor, and as the brilliant handmaid of the Sun comes further up, so the sky closes itself, light after light, even to the most beautiful."

Following al-Farghānī, Dante believed that the Earth's circumference was 20,400 Arab miles, not far from the real measure of 40,075 kilometers, given that the Arab mile equals 1973 meters. So six thousand miles correspond, if measured along the equator, to an arc of 106° in longitude. Since the astronomical twilight begins when the Sun is 18° below the horizon and since the length of the arcs along the parallels decrease at our latitudes with respect to what they are at the equator, we see that the distance of 6000 miles estimated by Dante between a place where the aurora begins and the one where the Sun is at noon ($90° + 18° = 108°$) is quite accurate.

7.5 The lightnings

At the time the *Comedy* was written, it was thought that lightnings developed in the regions where the sphere of air bordered that of fire, whose heat accumulated in the clouds until it was discharged with a flash. In canto XXXII of the *Purgatorio* Dante compared the lightning to the imperial eagle that descended very quickly on the chariot of the Church:

Non scese mai con sì veloce moto
foco di spessa nube, quando piove
da quel confine che più va remoto,
com'io vidi calar l'uccel di Giove
(*Pur.* XXXII, 109–112)

"Never has fire from a dense cloud descended with such swift motion, raining from the boundary that moves most remote from us, as I saw Jove's bird fall."

Later in canto XXIII of *Paradiso* he used the phenomenon of the lightning to indicate an action out of the ordinary:

Come foco di nube si diserra
per dilatarsi sì che non vi cape,
e fuor di sua natura in giù s'atterra,
(*Par.* XXIII, 40–42)

"As fire is unlocked from a cloud, expanding beyond containment there, and against its nature downwards unearths itself."

In fact, the fire of a lightning by its nature would tend to rise towards the sphere of fire (*Par.* I, 115), but the excessive accumulation in the cloud forces it to descend towards the ground. The same unnatural tendency of the lightning is also mentioned in canto I of the *Paradiso*:

e sì come veder si può cadere
foco di nube, sì l'impeto primo
l'atterra torto da falso piacere.
(*Par.* I, 133–135)

"and, just as one can see fire fall downward from a cloud, so the creature's first impetus drive it to Earth, if deflected by false pleasure."

7.6 The shooting stars

In the *Comedy* Dante referred to shooting stars in Aristotelian terms, describing them as vapors exhaled from the Earth that rising upwards burn due to the heat of the Sun. In canto V of the *Purgatorio*, to describe the speed with which the two messengers returned to report about Dante to the host of souls in the Antipurgatory, they were assimilated to the shooting stars at the beginning of the night and to the lightning in summer afternoons:

Vapori accesi non vid'io sì tosto
di prima notte mai fender sereno,
né, sol calando, nuvole d'agosto,
*   che color non tornasser suso in meno;*
(Pur. V, 37–40)

"I have never seen flaming vapors in early night render the clear sky or the clouds in August at sunset, and swiftly as those two went back up."

Later, in canto XV of the *Paradiso*, the poet compared his ancestor Cacciaguida to a shooting star:

Quale per li seren tranquilli e puri
discorre ad ora ad or sùbito foco,
movendo li occhi che stavan sicuri,
*   e pare stella che tramuti loco,*
se non che da la parte ond'e' s'accende
nulla sen perde, ed esso dura poco:
(Par. XV, 13–18)

"As across a cloudless, pure, and tranquil sky from time to time there darts a sudden fire, moving our eyes, previously fixed, and it seems a star changing place where it began no star is lost, and it lasts but little."

Here the shooting star, instead of steam ignited in the upper atmosphere, looks like a moving star; but it is not, since no star is missing and the phenomenon does not last long, unlike the fixed stars.

7.7   The comets

Only once in the *Commedia* Dante spoke of comets, when in canto XXIV of the *Paradiso* he described the apostles revolving in the sky:

[...] e quelle anime liete
si fero spere sopra fissi poli,
fiammando, volte, a guisa di comete.
(Par. XXIV, 10–12)

"and those happy souls became spheres spinning on fixed poles, flaming, as they turned, like comets."

In those days, comets were believed to be dry exhalations that rose from the sphere of air to that of fire, where they became visible.

Perhaps Dante did not like to talk about comets because they were considered to be carriers of misfortune: he knew it well, because, shortly before his two sentences to exile and death, in the autumn of 1301 Halley's

comet had been clearly visible in its passage near the Sun, which occurs
approximately every seventy-six years. Dante mentioned it in the *Convivio*
(**II**, XIII, 22):

> [...] *in Fiorenza, nel principio della sua destruzione, veduta fue
> nell'aere, in figura d'una croce, grande quantità di questi vapori
> seguaci della stella di Marte.*

"in Florence, at the beginning of its destruction, was seen in the air, in the
figure of a cross, a large quantity of these vapors, followers of the star of
Mars."

At that time Charles of Valois, who had been sent to Florence to pacify
it, instead witnessed without intervening in the taking of the city by the
Black Guelphs of Corso Donati, who put it to fire and sword and started
the persecution of the White Guelphs. This led, among other things, to
the two condemnations to Dante's exile and death, decreed on January 27
and March 10, 1302 by the *podestà* Cante Gabrielli of Gubbio, appointed
precisely by Valois, while the poet was withheld in Rome by Boniface VIII.

That same appearance of Halley's comet inspired Giotto, a contemporary of Dante, who painted it in the Adoration of the Magi in the Scrovegni
chapel in Padua.

The last appearance of Halley's comet was in 1986; in that passage
it was reached by the European probe Giotto, which photographed the
nucleus with an instrument largely built in Italy.

Chapter 8

The motions of the heavens

For Dante, the Earth is stationary at the center of the Universe. The celestial spheres rotate around it, each contiguous to the next. Immediately above the clouds (therefore above the Earthly Paradise, on top of the mountain of Purgatory) there is the sphere of fire, then the spheres of the nine heavens: Moon, Mercury, Venus, Sun, Mars, Jupiter, Saturn, the Fixed Stars and the Prime Mover; above the Empyrean (Fig. 6.2).

Each celestial sphere has two motions: one diurnal from east to west (rotation), with one turn every 24 hours to compensate for the fact that in reality the Earth is not stationary, but rotates around its axis, and one of a longer period (revolution) in the opposite direction, which corresponds to the real motions of the various celestial bodies. Dante mentioned these two motions at the beginning of canto X of the *Paradiso*:

> *Leva dunque, lettore, a l'alte rote*
> *meco la vista, dritto a quella parte*
> *dove l'un moto e l'altro si percuote;*
> (*Par.* X, 7–9)

"Lift therefore your gaze to the high wheels with me, reader, straight to that place where the one and the other motion strike each other."

The two motions meet at the two equinoctial points, the point γ and the point ω, where in fact the celestial equator and the ecliptic cross; these are the two maximum circles along which these motions take place. Furthermore, while the rotation takes place around the polar axis of the Earth, the motions of revolution are different for the various heavens: that of the Sun occurs around the axis of the ecliptic (i.e. the plane of the Earth's orbit around the Sun), which forms an angle of 23.5° with the polar axis of the Earth; that of the Moon around an axis close to the Earth's polar one; those of Mercury, Venus, Mars, Jupiter, and Saturn around axes close to that of the ecliptic. It is precisely the obliquity of the ecliptic that creates the seasons, which are the origin of life and order on Earth, as Dante noted in the continuation of canto X of the *Paradiso*:

> *Vedi come da indi si dirama*
> *l'oblico cerchio che i pianeti porta,*
> *per sodisfare al mondo che li chiama.*
> *Che se la strada lor non fosse torta,*
> *molta virtù nel ciel sarebbe in vano,*
> *e quasi ogne potenza qua giù morta;*
> *e se dal dritto più o men lontano*
> *fosse 'l partire, assai sarebbe manco*
> *e giù e sù de l'ordine mondano.*
> (*Par.* X, 13–21)

"See branching off from there the oblique circle that carries the planets, so as to satisfy the world that calls for them: for if their path were not twisted, much of the power in the heavens would be in vain, and dead almost every potentiality down here, and if its departure from the straight were greater or smaller, much would be lacking, both below and above, in the order of the world."

If the obliquity of the ecliptic were different, there would be serious consequences in the order of things in the northern and southern hemispheres.

Furthermore, while the Prime Mover only performs the rotational motion around the Earth's polar axis — and it is the one that moves most rapidly, dragging the other heavens, which are slower as one proceeds inward —, the heaven of the fixed stars, in addition to rotation, has a very slow revolution, also from east to west, around the axis of the ecliptic, with a period estimated at the time of 36,000 years. This corresponds to the precession motion, whose real period is instead 25,772 years. Dante mentioned this motion of the sky of the fixed stars in canto XI of the *Purgatorio*, saying that it was the slowest motion of the heavens: "*al cerchio che più tardi in cielo è torto*" (v. 108) "to the circle that turns slowest in the sky", and demonstrated that he knew it in canto 1 of the *Purgatorio* when he spoke of the four stars (see Section 11.1).

Chapter 9

The Moon

The heaven of the Moon is the closest to the Earth and the one with the slowest rotational motion (*"la spera più tarda"*, *Par.* III, 51, "the slowest sphere"). Dante and Beatrice crossed the sphere of fire at the speed of a lightning, reaching the heaven of the Moon in the time it takes for the arrow shot to hit the target. These are expedients through which Dante made realistic operations which would otherwise be dangerous and very long; like in a science fiction movie.

The Moon, described as "shining, dense, solid, clear, like a diamond struck by the Sun" (*Par.* II, 32–33), boasts several references in the *Commedia* for its mythological importance (it is often personified in Diana, "daughter of Latona", as in *Par.* XXII, 139–141), for its peculiar astronomical characteristics, for its uses in the measurement of time, and for its changing appearance that has led to speculation on the possible correlations between the lunar astro and the earthly events.

9.1 The setting of the Moon

Even before ascending to the heaven of the Moon, Dante referred to it several times. For example, at the beginning of canto X of the *Purgatorio*, the poet talked about the setting of Moon and noted that "the hollow", that was the dark edge, set first, while the bright part of the Moon was still visible:

> *E questo fece i nostri passi scarsi,*
> *tanto che pria lo scemo de la luna*
> *rigiunse al letto suo per ricorcarsi,*
>   *che noi fossimo fuor di quella cruna;*
>   (*Pur.* X, 13–16)

"And this made our steps so slow that the hollow of the moon reached its bed to lie down again, before we came forth from that needle's eye."

In fact, since three days had passed from the full moon, the dark edge was facing west, in accordance with the rule of thumb "hump to the east waning moon, hump to the west crescent moon" (in Italian it rhymes: *gobba a levante luna calante, gobba a ponente luna crescente*), where the hump is the luminous edge.

9.2   The Moon at midnight

In canto XVIII of the *Purgatorio* Dante and Virgil were resting for the night, the fourth after that of the full moon during which the poet was lost in the dark forest. While they were waiting to enter the fourth ledge, they saw the Moon appearing almost at midnight, behind the mountain of Purgatory, which had hidden its rising from the sea a few hours earlier.

> *La luna, quasi a mezza notte tarda,*
> *facea le stelle a noi parer più rade,*
> *fatta com'un secchion che tuttor arda;*
> * e correa contra 'l ciel per quelle strade*
> *che 'l sole infiamma allor che quel da Roma*
> *tra 'Sardi e 'Corsi il vede quando cade.*
> *(Pur. XVIII, 76–81)*

"The Moon, delayed almost until midnight, now made the stars seem fewer, shaped like a copper bucket still on fire, and it was running against the heavens along those paths which the Sun enflames when the Roman sees it set between the Sardinians and the Corsicans."

The nocturnal astro is in the constellation of Scorpio, in the position where the Sun is seen from Rome when it is setting in the Strait of Bonifacio ("between the Sardinians and the Corsicans"), that is, in early October. The astronomical notation is very precise and also takes into account the error of the Julian calendar (see Section 5.2): it is possible that Dante in early October 1301 was in Rome, where he had been sent as ambassador of the White Guelphs, shortly before the arrival of Charles of Valois in Florence on November 1 of the same year.

The "accusation" that Dante made to the Moon of making the stars seem fewer is fully shared by us astronomers, since, when we are going to observe weak objects with a telescope, we try to avoid the presence of the Moon in the sky, that is, we prefer the nights close to the new moon. In fact, the so-called "dark nights" are very popular and those who schedule the observations only grant them to the best scientific proposals; it must be said that when the Moon is almost full, the domes of some telescopes (e.g. the wide-field ones), particularly sensitive to the background light, are not even opened.

9.3 The geomancers

At the end of the night described above, Dante at the beginning of canto XIX of the *Purgatorio* said:

> *Ne l'ora che non può 'l calor dïurno*
> *intepidar più 'l freddo de la luna,*
> *vinto da terra, e talor da Saturno*
> *– quando i geomanti lor Maggior Fortuna*
> *veggiono in orïente, innanzi a l'alba,*
> *surger per via che poco le sta bruna –,*
>   (*Pur.* XIX, 1–6)

"At the hour when the heat of the day can no longer warm the cold of the Moon, vanquished by Earth and sometimes by Saturn, when the geomancers see their Greater Fortune in the east, before the dawn, rising along the path that stays dark for it but a short while."

He painted the moment before dawn with a double paraphrase that reported the astronomical and geographical doctrines of the time: in the first he referred to the belief that currents of cold air spread over the Earth from the Moon and Saturn. It is interesting that for Dante it is the "heat of the day/[...] vanquished by Earth", that is, that the Earth also cools, thus approaching scientific truth, since obviously it is not the Moon that cools our planet.

In the following triplet Dante indicated the same hour when speaking of the geomancers (fortune tellers who randomly marked on the ground some points from which they derived lines and figures with various meanings) and of when they saw the rise of "*lor Maggior Fortuna*", one of the typical figures of geomancy. It was formed by points with an arrangement similar to that of the eastern half of Aquarius and the western half of Pisces, which rose just before dawn that day (March 29, 1301).

9.4 Penetration into the Moon

Dante and Beatrice at the beginning of canto II of *Paradise* entered the body of the Moon:

> *Parev'a me che nube ne coprisse*
> *lucida, spessa, solida e pulita,*
> *quasi adamante che lo sol ferisse.*
>   *Per entro sé l'etterna margarita*
> *ne ricevette, com'acqua recepe*
> *raggio di luce permanendo unita.*
>   *S'io era corpo, e qui non si concepe*
> *com'una dimensione altra patio,*
> *ch'esser convien se corpo in corpo repe,*
> *accender ne dovria più il disio*
> *di veder quella essenza in che si vede*
> *come nostra natura e Dio s'unio.*
>   (*Par.* II, 31–42)

"It seemed to me that a cloud covered us, shining, dense, solid, clear, like a diamond struck by the Sun. Within itself the eternal pearl received us, as water receives a ray of light while still remaining whole. If I was a body — and down here it cannot be conceived how one dimension could accept another, as must occur, if body coincide with body — it should kindle within us more desire to see that Essence where is seen how our nature and God became one."

Unlike Anaximander and Plato, who considered the Moon an igneous body with its own luminosity, Alighieri believed that it was "dense and solid" and that it reflected the light of the Sun, otherwise the solar eclipses, in which the Moon covers the Sun, could not be explained (see below).

Therefore Dante was able to penetrate the Moon, transgressing the principle of impenetrability of bodies, thanks to a supernatural phenomenon that assimilated his nature to that of God. Only in cases where his baggage of scientific knowledge did not offer him the tools to fulfil the goals of the trip, the poet resorted to the supernatural.

9.5 The Moon spots

Continuing in the reading of the canto of the *Paradiso*, still in the sky of the Moon, we find a beautiful exchange between Beatrice and Dante (vv. 49–111), which is worth following in detail because it is rich in astronomical and scientific topics; indeed, it even concerns the scientific method. The poet asked the woman what the Moon spots were, in which popular imagination saw the figure of Cain (as already underlined in canto XX of the *Inferno*):

> *Ma ditemi: che son li segni bui*
> *di questo corpo, che là giuso in terra*

fan di Cain favoleggiare altrui?"
   (*Par.* II, 49–51)

"But tell me: what are the dark marks in this body, that make people down there on Earth tell fables about Cain?"

Beatrice first warned Dante that sometimes, following what the senses tell us, our reason is unable to grasp the truth. Then she turned the question around and asked him what he thought about the Moon spots:

> *Ella sorrise alquanto, e poi "S'elli erra*
> *l'oppinïon", mi disse, "d'i mortali*
> *dove chiave di senso non diserra,*
>   *certo non ti dovrien punger li strali*
> *d'ammirazione omai, poi dietro ai sensi*
> *vedi che la ragione ha corte l'ali.*
>   *Ma dimmi quel che tu da te ne pensi".*
> (*Par.* II, 52–58)

"She smiled a little, 'If,' she said, 'the opinion of mortals errs where no key of sense unlocks, surely the arrow of wonder ought not to pierce you now, since you see that reason has short wings even when following the senses. But tell me what you think of it yourself.'"

The poet replied that he thought they were due to differences in density, referring to an idea of Averroes, accepted by Albert the Great (see also *Convivio* **II**, XIII, 9):

> *E io: "Ciò che n'appar qua sù diverso*
> *credo che fanno i corpi rari e densi".*
> (*Par.* II, 59–60)

"And I: 'What looks different to us up here is caused, I think, by bodies rare and dense.'"

Beatrice's reply began pompously with an introductory warning. Then she noted that the eighth sphere, the heaven of the fixed stars, showed many lights of varying intensity: if this were due to differences in density, the stars would have only one virtue, albeit variously distributed. Instead, each star has its own special virtue:

> *Ed ella: "Certo assai vedrai sommerso*
> *nel falso il creder tuo, se bene ascolti*
> *l'argomentar ch'io li farò avverso.*
>   *La spera ottava vi dimostra molti*
> *lumi, li quali e nel quale e nel quanto*

> *notar si possono di diversi volti.*
> *Se raro e denso ciò facesser tanto,*
> *una sola virtù sarebbe in tutti,*
> *più e men distribuita e altrettanto.*
> *Virtù diverse esser convegnon frutti*
> *di princìpi formali, e quei, for ch'uno,*
> *seguiterieno a tua ragion distrutti.*
> *(Par.* II, 61–72)

"And she: 'Certainly you will find your belief much submerged in error, if you listen carefully to the argument I shall make against it. The eight sphere displays to you many lights, which both in quality and size can be seen to have different faces. If rare and dense alone caused that, one sole power would be in all of them, distributed into more and less, and some times equally. Different powers must necessarily be the fruit of formal principles, and these, except for one, would according to your thinking be destroyed.'"

As Thomas Aquinas said (*In libros Aristotelis De caelo et mundo expositio*, II, lect. 19): "it is necessary that the highest sphere abounds in the quantity of stars in which the various active virtues are found" (see also *Questio de aqua et terra*, XXI, 70–71).

Then Beatrice gave another physical refutation of Dante's idea. If the Moon has less dense regions, either these do not cover the whole face of the Moon, or they do cover it. In the first case, parts of the Moon would be transparent and the confirmation of this hypothesis would occur during a solar eclipse, but this is not the case. In the second case the light, penetrating into the Moon, should at a certain point meet a denser area that does not let it pass, reflecting it like a mirror (the mirrors were made of glass covered with a layer of lead). And it would be useless to argue that in the darker areas the denser reflective layer is further back than in the light areas:

> *Ancor, se raro fosse di quel bruno*
> *cagion che tu dimandi, o d'oltre in parte*
> *fora di sua materia sì digiuno*
> *esto pianeto, o, sì come comparte*
> *lo grasso e 'l magro un corpo, così questo*
> *nel suo volume cangerebbe carte.*
> *Se 'l primo fosse, fora manifesto*
> *ne l'eclissi del sol, per trasparere*
> *lo lume come in altro raro ingesto.*
> *Questo non è: però è da vedere*

de l'altro; e s'elli avvien ch'io l'altro cassi,
falsificato fia lo tuo parere.
  S'elli è che questo raro non trapassi,
esser conviene un termine da onde
lo suo contrario più passar non lassi;
  e indi l'altrui raggio si rifonde
così come color torna per vetro
lo quale di retro a sé piombo nasconde.
  Or dirai tu ch'el si dimostra tetro
ivi lo raggio più che in altre parti,
per esser lì refratto più a retro.
    (*Par.* II, 73–93)

"Again, if rare matter were the cause of darkness you inquire about, either the rare matter would go entirely through, where there are spots in this planet, or, as a living body alternates fat and lean, so it would change pages through its volume. The first case would be manifested in eclipses of the Sun, for the light would shine through, as it does through any rare matter. This is not so: therefore let us look to the other possibility, and, if I break that down, too, your opinion would be shown to be false. If the rare matter is not continuous, there must be a limit where its contrary blocks passage, and from there, rays of light must bound back, just as color returns through glass that has lead hidden behind it. Now you will say that the ray appears darker there than elsewhere because it is reflected from further back."

To refute the possible objection, Beatrice resorted to a "thought experiment", a sort of Einstein-style *Gedankenexperiment*. Get three mirrors; you put two of them at an equal distance from you, the third farthest and in the middle of the other two. Putting a lamp behind yourself, you will see that it shines equally in all the mirrors, even if in the farthest one it appears smaller. This observation anticipates the theory of the independence of surface brightness from distance, according to which the light emitted per unit of solid angle from a light source does not depend on its distance. The reason is that the surface brightness is the ratio between the luminosity of the body and the solid angle that the body covers: since both quantities are inversely proportional to the square of the distance, their ratio does not change with distance (unless there is absorption of the light by some medium interposed between source and observer).

Da questa instanza può deliberarti
esperïenza, se già mai la provi,

ch'esser suol fonte ai rivi di vostr'arti.
Tre specchi prenderai; e i due rimovi
da te d'un modo, e l'altro, più rimosso,
tr'ambo li primi li occhi tuoi ritrovi.
Rivolto ad essi, fa che dopo il dosso
ti stea un lume che i tre specchi accenda
e torni a te da tutti ripercosso.
Ben che nel quanto tanto non si stenda
la vista più lontana, lì vedrai
come convien ch'igualmente risplenda.
(*Par.* II, 94–105)

"From this objection an experiment can free you, if you ever try it, for from experience derive the streams of all your arts. Take three mirrors, and place two of them at the same distance from you, and let your eyes find the third more distant and between the first two. Facing toward them, have a light from behind you shine on the three mirrors and return to you reflected from all three. Even though the more distant image is not as extended in size, you will see that it is equally bright there."

Finally Beatrice, after having exhaustively refuted Dante's hypothesis, in the rest of the canto enunciated what would be the true cause of the Moon spots: they depended on the virtue that from the Empyrean was variously distributed to the skies below.

The question of the lunar spots is ancient and subtle. Aristotle had postulated that the Moon, like all other wandering stars, was a homogeneous sphere composed of ether, a mysterious substance proper to celestial bodies. An idea completely different from that of Anaxagoras, a Greek philosopher active in Athens in the V century BC, who instead theorized a substantial structural homogeneity between the terrestrial and the celestial worlds. He also considered the Moon a large stone without its own light, unlike the other stars that he imagined as fiery stones, and explained the Moon spots as the shadows caused by the wrinkles of its surface. The question became a subject of heated debate in the Middle Ages, despite the constant appeal to the *auctoritates* that could make digestible even the most bitter morsels for reason. Someone even revived Anaxagoras' proposal, which had already cost the author a sentence for impiety. However most scholars thought of overcoming the impasse by postulating that the spots were the consequence of excesses and defects in the density of the ether, skirting that even this solution violated the Aristotelian postulate of the perfection of the ether.

Finally, three centuries after the writing of the *Commedia*, Galileo Galilei, observing our satellite with the telescope, would have written about it in the *Sidereus Nuncius*:

> [...] *della faccia lunare che è rivolta al nostro sguardo* [...] *io distinguo in due parti, più chiara e più oscura. La più chiara par circondare e cosparger di sè tutto l'emisfero; la più scura invece, offusca a guisa di nuvola la faccia stessa e la fa apparire macchiata. Ora queste macchie, alquanto oscure e abbastanza ampie, sono visibili ad ognuno, e sempre in ogni epoca furono scorte; e perciò le chiameremo grandi, o antiche, a differenza di altre macchie, minori per ampiezza, ma così fitte, da ricoprire tutta la superficie lunare, e specialmente la parte più lucente. Queste invero da nessuno furono osservate prima di noi* [...].

"[...] of the lunar face that is turned to our sight [...] I distinguish in two parts, lighter and darker. The lighter seems to surround and sprinkle with itself the whole hemisphere; the darker one, on the other hand, obscures the face itself like a cloud and makes it appear spotted. Now these spots, somewhat obscure and rather large, are visible to everyone, and they were always seen in every age; and therefore we will call them large, or ancient, unlike other spots, smaller in size, but so dense, as to cover the entire lunar surface, and especially the brightest part. These were indeed observed by no one before us [...]."

To conclude this discussion on the origin of the Moon spots, we would like to highlight the fact that Dante anticipated the scientific method by making use of real or thought experiments and anticipated the theory of the independence of surface brightness from distance (for example, it is fundamental in the formulation of Olbers' paradox that questions how the night sky can be dark by making certain assumptions such as infinity of the cosmos in space and time, along with homogeneity and isotropy, which are thus falsified). Furthermore Dante elects a woman as his scientific teacher; it would then take many centuries before women could be accepted in science: Dante had already done it, well in advance.

9.6   A lunar clock

The Moon with its phases is an efficient analog clock, used by many peoples (including Jews and Chinese) in synergy with the Sun as a metronome for the calendar. Even Christians, whose calendar is based entirely on the annual motion of the Sun, still use the inconstant Selene to fix the date of the movable feast of Easter and as an approximate measure of time over short periods. A well-known example is found in the verses of the *Inferno* in which Farinata degli Uberti, the "enemy" that Dante respected and admired, because after the battle of Montaperti he was the only one, "there where all others would have suffered Florence to be razed", to defend the city "openly" (*Inf.* X, 91–93), prophesied his exile in about four years:

> Ma non cinquanta volte fia raccesa
> la faccia de la donna che qui regge,
> che tu saprai quanto quell'arte pesa.
>   (*Inf.* X, 79–81)

"But not fifty times will be rekindled the face of the lady who reigns here, before you will know much that art weighs."

The lady is the Moon, identified with Proserpina-Ecate, Pluto's wife and queen of the underworld.

In canto XXVI of the *Inferno* (vv. 130–132), while Ulysses told his journey beyond the columns of Hercules that led him to see the mountain of Purgatory, the Moon was called again into question to measure the duration of the long wandering in pursuit of "virtue and knowledge", which lasted for five lunar months (see Section 6.2).

The Moon came back once more in the Cocito, when Count Ugolino, "speaking and weeping", told Dante about the premonitory dream of his misfortunes, which occurred after a few months of imprisonment during which he had seen the phases of the Moon making several turns from a small window of the Muda tower, where Archbishop Ruggieri had relegated him:

> m'avea mostrato per lo suo forame
> più lune già, quand'io feci 'l mal sonno
> che del futuro mi squarciò 'l velame.
>   (*Inf.* XXXIII, 25–27)

"had shown me through its opening several moons already, when I dreamed the evil dream that rent the veil of the future for me."

And again, when Virgil warned him that "Already the Moon is beneath our feet" (*Inf.* XXIX, 10), to signify that it was late and there was no need to linger further in the ninth pit of the eighth circle, among the sowers of discord, where the poet sought Geri del Bello, his distant relative.

More articulated is the passage of the *Paradiso* in which Dante, to represent a very brief pause of silence by Beatrice after he had been dazzled by the divine light, used a very complicated astronomical metaphor. The poet referred to the Sun and the Moon, identified with Apollo and Diana, children of Latona, when on the occasion of the vernal equinox, one in Aries and the other in Libra because they were occasionally located on opposite sides with respect to the zenith ("cenìt" in Dante's spelling), in the blink of an eye they crossed the celestial equator, the Sun to move to the northern hemisphere and the Moon to the southern one:

> *Quando ambedue li figli di Latona*
> *coperti del Montone e de la Libra,*
> *fanno de l'orizzonte insieme zona,*
>   *quant'è dal punto che 'l cenìt inlibra,*
> *infin che l'uno e l'altro da quel cinto,*
> *cambiando l'emisperio, si dilibra,*
>   *tanto, col volto di riso dipinto,*
> *si tacque Bëatrice, riguardando*
> *fiso nel punto che m'avëa vinto.*
> (*Par.* XXIX, 1–9)

"When both Latona's children, covered by the Ram and by the Scales, together make the horizon their girdle, as long a time as from the point when the zenith has them balanced until both slide from equilibrium on that belt, changing hemispheres: so long, her face covered with a smile, was Beatrice silent, gazing fixedly into the point that had vanquished me."

Dante also made frequent references to the phases of the Moon, as metaphors of an hour, of a time, or simply of a condition of light. As when, in canto XV of the *Inferno* (vv. 18–19) he and Virgil met the crowd of the sodomites (among which the poet would recognize his teacher, Brunetto Latini), who gazed at them in the manner of those "in the evening / looking at one another under new moon", that is, when it is pitch dark.

9.7 The tides

In canto XVI of the *Paradiso*, the ancestor Cacciaguida represented to Dante the alternating fortunes of his beloved Florence using the simile of fate that behaved like the Moon with the tides:

> *E come 'l volger del ciel de la luna*
> *cuopre e discuopre i liti sanza posa,*
> *così fa di Fiorenza la Fortuna:*
> (*Par.* XVI, 82–84)

"And as the turning of the heaven of the moon covers and uncovers the shores without pause, so Fortune does with Florence."

The phenomenon of the tides was also mentioned in canto XV of the *Inferno*, where the poet made one of his frequent geographical references, in this case to the embankments erected by the Flemings to counter the onslaught of the waters between Wissant and Bruges (the names of the places are Italianized to create an onomatopoeia that calls fire):

> *Quali Fiamminghi tra Guizzante e Bruggia,*
> *temendo 'l fiotto che 'nver' lor s'avventa,*
> *fanno lo schermo perché 'l mar si fuggia;*
> (*Inf.* XV, 4–6)

"As the Flemings, between Wissant and Bruges, fearing the tide that rises against them, make dikes to escape the sea."

Throughout history, men have tried to account for the phenomenon of the tides using the weapons of both the wildest imagination and reason: the solutions gradually proposed appear sometimes extravagant (like the Japanese idea of a god of the sea who controls the periodic rising of the waters through a magical jewel), but often correctly refer to the Moon, albeit in ignorance of the true reasons for the role of the star Diana.

In the Mediterranean basin, where the tides have modest amplitudes, the correlation between the rising and setting of the Moon and the fluctuating trend of the waters (which rise and fall approximately twice a day) had been observed since the IV century BC. Almost ignored by Aristotle, the phenomenon was examined by the Greek philosopher, geographer, and historian Posidonius of Rhodes, who lived between the II and I centuries BC. He attributed it to some unknown influence of the stars, especially of the Moon: an idea later revived in astrological key by Ptolemy.

At the time of Dante there were theories of a vitalistic nature, both of classical and Arab origin, which assigned to some function of the Earth organism the task of periodically modulating the height of the waters. It seems that the great poet, however, had not received them, given his peremptory association of the phenomenon of the tides to the Moon. Another masterful intuition that, after Galilei's unsuccessful attempt to explain the ebb and flow of waters through the rotation and revolution of the Earth, would have found full physical justification with Isaac Newton over three and a half centuries later.

It is perhaps interesting to note that even then the greater "power" of the Sun, now king of the solar system, was clear compared to the dimmer

Moon, which instead seemed to dictate the rules of the game with regard to the waters. The fact is that, Newton explained, the tides do not respond to gravity tout-court, but to its variation along the extension of the attracted body, and in this game the overwhelming mass of the Sun is penalized by its much larger distance than the Moon.

Chapter 10

The planets

10.1 Mercury

Dante and Beatrice ascended "to the second kingdom", the second heaven, the sphere of Mercury, where they met the Byzantine emperor Justinian. Dante asked him why he was in that sphere "that veils itself from mortals with another's rays" (*Par.* V, 129), that is, with the rays of the Sun almost always blocking the view. In fact Mercury, being the planet closest to the star and having a small orbit inside that of the Earth, always remains very close to the Sun (within 28°). Justinian replied that the small celestial body hosted those who did good in life but, because they were driven by the desire for honor and fame, deserved to be further away from God:

> *Questa picciola stella si correda*
> *d'i buoni spirti che son stati attivi*
> *perché onore e fama li succeda:*
> (*Par.* VI, 112–114)

"The little star is adorned by good spirits who were active so that honor and fame might follow them."

10.2 Venus

The next heaven is the one of Venus, which takes its name from the goddess of the "mad desire".

> *Solea creder lo mondo in suo periclo*
> *che la bella Ciprigna il folle amore*
> *raggiasse, volta nel terzo epiciclo;*
> (*Par.* VIII, 1–3)

117

"The world used to believe, to its peril, that the lovely Cyprian radiated mad desire, turning the third epicycle."

The planet Venus also has an orbit inside that of the Earth and therefore remains close to the Sun (within 48°), sometimes preceding it, sometimes following it ("the star that woos the Sun, now at his nape, now at his brow", *Par.* VIII, 11–12). Therefore Venus can be seen either in the west just after sunset, or in the east just before sunrise. Like the Moon, Venus is also mentioned several times in the *Commedia*, even before Dante and Beatrice reached its heaven. Let's see some passages.

10.2.1   *The lovely planet that strengthens us to love*

At the beginning of the *Purgatorio*, Dante had just come out of Hell "to look again at the stars" and, after invoking the muse of poetry Calliope, raised his eyes to heaven just before dawn; the first star he saw was Venus:

> Lo bel pianeto che d'amar conforta
> faceva tutto rider l'orïente,
> velando i Pesci ch'erano in sua scorta.
>   (*Pur.* I, 19–21)

"The lovely planet that strengthens us to love was causing all the east to laugh, vailing the Fish, which were her escort."

Indeed Venus is very easily seen, since it is the brightest object in the sky at night after the Moon. As we noted in Section 5.1, this step is fundamental for the dating of Dante's journey, as in the spring of 1301 Venus was visible to the east in the morning and was in the constellation of Pisces, while in the spring of 1300 it was visible to the west, in the evening, and in Taurus.

10.2.2   *The rise of Citerea*

In canto XXVII of the *Purgatorio*, just before dawn on March 30, 1301, Dante had the third premonitory dream and before he saw Venus again:

> Ne l'ora, credo, che de l'orïente
> prima raggiò nel monte Citerea,
> che di foco d'amor par sempre ardente,
>   (*Pur.* XXVII, 94–96)

"At the hour, I believe, when from the east Citerea first shone on the mountain, she who seems always aflame with the fire of love."

Venus (the goddess was very "venerated" on the island of Kythera) that day rose in Purgatory around three in the morning and it was precisely a morning star, further confirming the dating of the trip to 1301.

10.2.3 The shadow cone of the Earth

In canto IX of the *Paradiso* Dante said that the tip of the shadow cone of the Earth reached up to the sky of Venus:

> *Da questo cielo, in cui l'ombra s'appunta*
> *che 'l vostro mondo face [...]*
> (*Par.* IX, 118–119)

"Into this heaven, where the shadow of your world shrinks to a point..."

This note is exact according to the values given by the Persian astronomer al-Farghānī for the distances and diameters of celestial bodies. In fact, since the Earth's orbit around the Sun is external to that of Venus, its shadow goes in the opposite direction, therefore it cannot reach Venus.

It is interesting to note that the measures of the solar system adopted by the ancients, including those just mentioned, could have been disproved. In fact, if the shadow cone of the Earth had reached Venus, eclipses produced on Venus by the Earth would have had to be observed. However, it must be said that the decrease in luminosity of Venus caused by the Earth would have been very small, therefore not observable with the naked eye.

10.3 Sun

The Sun in the Ptolemaic system was one of the seven stars, called planets, revolving around the Earth. In fact, in canto I (v. 17) of the *Inferno* Dante called the Sun "planet", that is, etymologically "wandering star"; a concept that was reaffirmed in canto XXII of the *Paradiso* (v. 148).

Rising to the heaven of the Sun, Dante said:

> *Lo ministro maggior de la natura,*
> *che del valor del ciel lo mondo imprenta*
> *e col suo lume il tempo ne misura,*
> *con quella parte che sù si rammenta*
> *congiunto, si girava per le spire*
> *in che più tosto ognora s'appresenta;*
> (*Par.* X, 28–33)

"The greatest minister of nature, which stamps the world with the power of the heavens and measures time to us with its light, joined with that part mentioned above, was wheeling through the spirals in which he rises ever earlier."

Indeed, the Sun marks the measure of time, both with its diurnal motion along the parallels, which determines hours and days, and with its annual motion along the ecliptic, which creates the alternation of the seasons. The combination of the two motions generates a spiral that rises to the north during winter and spring and returns to the south during summer and autumn.

10.3.1 *The Sun in the sign of Aries*

At the beginning of the *Inferno* Dante, when he was still outdoors, made this observation:

> *Temp'era dal principio del mattino,*
> *e 'l sol montava 'n sù con quelle stelle*
> *ch'eran con lui quando l'amor divino*
> *mosse di prima quelle cose belle;*
> (*Inf.* I, 37–40)

"The time was the beginning of the morning, and the Sun was mounting up with those stars that were with it when God's love first set those lovely things in motion."

That morning (which we believe to be that of March 25, 1301; see Chapter 2) the Sun was with those stars with which it was at the time of creation. It was a general opinion in antiquity and in the Middle Ages that the creation of the world took place in spring. Virgil also affirmed it:

> *Non alios prima crescentis origine mundi*
> *inluxisse dies aliumve habuisse tenorem*
> *crediderim; ver illud erat, ver magnus agebat*
> *orbis et hibernis parcebant flatibus euri,*
> *cum primae lucem pecudes haussere virumque*
> *terrea progenies duris caput extulit arvis*
> *immissaeque ferae silvis et sidera caelo.*
> (*Georgiche*, II, 336–342)

"I could believe that not different were the days that shone at the first dawn of the nascent world, or of a different climate: spring was that, spring the great Universe expired and the sirocco held his winter breath, when the first animals absorbed the light and men, terrestrial lineage, they pushed their heads out of the crust of the arable fields and the wild beasts were placed in the forests and the stars in the sky."

In 1301 the vernal equinox took place in the evening of March 12, earlier than March 21 due to the error of the Julian calendar, which Dante knew (see Section 5.2). So in the morning of March 25, 1301, it was the thirteenth day of spring and the Sun was in the sign of Aries, just over 12° from the γ point (the meeting point between the celestial equator and the ecliptic). In fact, the Sun, as seen from the Earth that turns around it, travels along the ecliptic about 1° per day, since it makes a full turn in 365 days. The period of the zodiac sign of Aries is set between 21 March and 20 April, because the Sun was actually in the constellation of Aries in that fraction of the year in the III–II century BC, when the zodiac signs were fixed. However, due to the precession of the equinoxes, the γ point shifts west by 50″ per year. So already at the time of Dante it was no longer in the constellation of Aries, but in that of Pisces, where it still is.

The problem is that at the time of the creation of the world, which was placed 6500 years before Dante's journey (see Section 5.3), the Sun was neither in the constellation of Pisces nor in that of Aries, but in that of Gemini. For this reason it is not possible that on March 25, 1301, the Sun was in the same position in the sky ("with those stars") in which it was at the moment of creation, in the spring of 6500 years earlier.

The only way not to believe that Dante made a mistake is that by "those stars" he meant that "zodiac sign": in fact, the zodiac sign of spring is always Aries and remains fixed, even though the γ point moves.

10.3.2 The Sun under Aquarius

At the beginning of canto XXIV of the *Inferno* Dante described his state of mind first desperate, then resigned, assimilating it to that of a young shepherd who on a winter day was discouraged seeing the meadows covered with frost and then calmed himself by seeing the frost melt:

> *In quella parte del giovanetto anno*
> *che 'l sole i crin sotto l'Aquario tempra*
> *e già le notti al mezzo dì sen vanno,*
> (*Inf.* XXIV, 1–3)

"In that part of the youthful year when the Sun tempers its locks under Aquarius and already the nights are moving south."

The Sun is in the sign of Aquarius (not in the constellation of the same name; see above) from January 21 to February 20 and in that period the nights are already getting shorter; their duration is approaching half of twenty-four hours, which they will reach at the vernal equinox on March 21. The astronomical situation, with an ever shorter duration of the hours of darkness, suggests the idea of calming the previously worried soul.

10.3.3   *The Sun and its symbols*

Like the Moon, the Sun also recurs frequently in Dante's works in its role as a celestial body responsible for various astronomical phenomena, but also as a synonym of light, wisdom, and creative power: hence the symbolic use as an image of the Virgin Mary, the Church, the pontiff, and the emperor. Some examples will help us understand.

We find the Sun as an allegory of the day that is born in the canto XXIII of the *Paradise*. This opens with the metaphor of the little bird that "with ardent affection the Sun waits, / fixedly looking for the dawn to be born" (vv. 8–9), anxiously awaiting that light which, having swept away the dangers of the night, allows him to go in search of food for his young.

The reference to the diurnal journey of the star in canto XXXIV of the *Inferno* is more subtle, where Dante asked himself: "and how, in so little time, has the Sun made the passage from evening to morning?" (vv. 104–105). The poet intended to give the reader an account of a sudden jump of twelve hours (see Section 5.1) and to this end he used a rhetorical device: he pretended not to understand why, after having crossed the center of the Earth, Lucifer's icy cage with Virgil, and having risen to the surface (of a spherical Earth) in the belief that he could "look again at the stars" (in fact it was afternoon in the northern hemisphere), with (fake) surprise he found himself in the early morning. What happened? The two poets had diametrically crossed the Earth, suddenly changing hemisphere, and therefore moving into the opposite half-day to the one they started from, which involved the shift of twelve hours and therefore the transition from evening to morning.

The Sun is often used to indicate a direction: for example the west, as in the verses in which Ulysses, urging his companions to dare in order to satisfy the thirst for knowledge, spurred them to dive into the unknown beyond the Pillars of Hercules in the wake of the star: "do not deny the experience, following the Sun, of the world without people" (*Inf.* XXVI, 116–117).

Dante made extensive use of the most obvious properties of the Sun, such as its ability to produce light and heat, using them in a physical sense but also as metaphors of the power of God; for example, the inability of men to look at the star and at the same time the anxiety to grasp its light:

> *Ma come al sol che nostra vista grava*
> *e per soverchio sua figura vela,*
> *così la mia virtù quivi mancava.*
> (*Pur.* XVII, 52–54)

"But as if the Sun, which weighs down our gaze and veils its shape with excess, so my power failed before him."

In these verses, the gaze of Dante, who had just recovered from one of the usual fainting due to an excess of emotion, was overwhelmed by the radiance of the angel of meekness, who had come to exhort him and Virgil to continue on the climb.

Formulated by Plato in the *Republic*, the equation God = light finds various applications throughout Dante's works. In fact it has very ancient precursors. Man has always shown a natural repulsion for darkness, identifying instead the very principles of life in light and heat. Just think of Eastern religions, Zoroastrianism and the sparkling Greek gods. Even the biblical God shines with intense light: this does not represent just a decoration, but is rather the symbolic underlining of an unattainable spirituality. For Dante, light is a sensitive manifestation of the presence of God. In this way Dante's metaphors should be read: from the "dark forest" of canto I of the *Inferno* to the qualification of the Empyrean as the "heaven that receives most of his light" (*Par.* I, 4).

10.4 Mars

In canto XIV of the *Paradiso* Dante ascended to the heaven of Mars, the red planet, which welcomed him by becoming redder:

> *Ben m'accors'io ch'io era più levato,*
> *per l'affocato riso de la stella,*
> *che mi parea più roggio che l'usato.*
> (*Par.* XIV, 85–87)

"I well perceived that I had risen higher, by the fiery smile of the star, which seemed to me more flame-colored than usual."

The reddening of Mars setting over the sea had already been mentioned in a simile in canto II of the *Purgatorio*:

> *Ed ecco, qual, sorpreso dal mattino,*
> *per li grossi vapor Marte rosseggia*
> *giù nel ponente sovra 'l suol marino:*
> (*Pur.* II, 13–15)

"And behold, as when near the morning Mars shines red through the heavy vapors, low in the west, over the surface of the sea."

10.4.1 *Mars in Leo*

In canto XVI of the *Paradiso* Cacciaguida indicated his date of birth with the number of Martian years (580) that have passed since the incarnation of Christ, counting the passages of the planet in Leo:

> *al suo Leon cinquecento cinquanta*
> *e trenta fiate venne questo foco*
> *a rinfiammarsi sotto la sua pianta.*
> (*Par.* XVI, 37–39)

"this fire came five hundred, fifty, and thirty times to its Lion, to be inflamed again beneath its foot."

The period of revolution of Mars given by the *Almagest* is 687 days, so Cacciaguida's year of birth is 1091 ($= 580 \times 687/365$). Mars was actually in Leo in the spring of 1301 during Dante's journey (while it was in Pisces in the spring of 1300), providing further confirmation of the dating we adopted (see Section 5.1).

10.5 Jupiter

Dante in canto XVIII of the *Paradiso* realized that he had risen to the heaven of Jupiter (the sixth) "seeing that miracle grows lovelier" (v. 63), where the miracle was obviously Beatrice, "in the whiteness of the temperate sixth star" (vv. 68–69), and to do so he took the time required by a woman's face to regain its normal color after blushing. With this he also alluded to the passage from the blush of Mars to the whiteness of Jupiter.

10.6 Saturn

In canto XXI of the *Paradiso* Dante, ascending to the heaven of Saturn, said that it shone "under the breast of the burning Lion" (v. 14). Indeed, in the spring of 1301 Saturn was in Leo and precisely near the star Regulus, which represented the chest of the animal (while in the spring of 1300 it was between Leo and Cancer).

In the *Convivio*, the sky of Saturn was associated with astronomy:

> *E lo cielo di Saturno hae due propietadi per le quali si può com-*
> *parare all'Astrologia: l'una si è la tardezza del suo movimento per*
> *[li] dodici segni, ché ventinove anni e più, secondo le scritture delli*
> *astrologi, vuole di tempo lo suo cerchio; l'altra si è che sopra tutti*
> *li altri pianeti esso è alto.*
>
> *E queste due propietadi sono nell'Astrologia: chè nel suo cerchio*
> *compiere, cioè nello aprendimento di quella, volge grandissimo*
> *spazio di tempo, sì per le sue [dimostrazioni], che sono più che*
> *d'alcuna delle sopra dette scienze, sì per la esperienza che a bene*
> *giudicare in essa si conviene*
> *[. . .].*
> *è altissima di tutte l'altre; però che, sì come dice Aristotle nel*
> *cominciamento dell'Anima, la scienza è alta di nobilitade per la*
> *nobilitade del suo subietto e per la sua certezza*
> *[. . .].*
> (*Convivio* **II**, XIII, 28–30)

"And the sky of Saturn has two properties for which it can be compared to astrology: one is the slowness of its movement for [the] twelve signs, since twenty-nine years and more, according to the writings of the astrologers, takes the time of his circle; the other is that it is high above all the other planets. And these two properties are in astrology: because in the fulfillment of its circle, that is, in the learning of that, it takes a very large space of time, both for its [demonstrations], which are more than in any of the aforementioned sciences, and for the experience which, if judged well, is convenient in it. [...] it is the highest of all the others; since, as Aristotle said in the beginning of his work *On the Soul*, science is high in nobility for the nobility of its subject and for its certainty [...]."

Then Dante continued with the exaltation of astronomy already reported in the Introduction.

Chapter 11

The stars

Dante mentioned the stars many times and mistakenly believed that they reflected the light of the Sun, like the Moon and the planets; in fact in canto XX of the *Paradiso* he said that after sunset:

> lo ciel, che sol di lui prima s'accende,
> subitamente si rifà parvente
> per molte luci, in che una risplende;
> (*Par.* XX, 4–6)

"the sky, which earlier was lit by him alone, suddenly becomes visible again with the many lights in which one light is reflected."

In canto XXII of the *Paradiso* the poet ascended to the heaven of the stars and reached it in the constellation of Gemini, which was his zodiac sign, from where, he affirmed, "I acknowledge that all my talent comes, whatever it may be" (vv. 113–114).

11.1 The four stars

In canto I of the *Purgatorio*, just out from Hell, Dante said:

> I' mi volsi a man destra, e puosi mente
> a l'altro polo, e vidi quattro stelle
> non viste mai fuor ch'a la prima gente.
> (*Pur.* I, 22–24)

"I turned to the right and considered the other pole, and I saw four stars never seen except by the first people."

It is likely that the four stars near the south pole refer to the constellation of the Southern Cross. They are not visible at our latitudes but become visible from southern Egypt. Known to Ptolemy, who had cataloged and incorporated them into the constellation of Centaurus, they continued to be relatively familiar to the Arabs. Dante should have known them, even if he had never observed them.

It is more difficult to understand who the "first people" are. It may be that Alighieri referred to Adam and Eve and to the fact that the Earthly Paradise was on the top of the mountain of Purgatory in the southern hemisphere. However, it is unlikely that he spoke of "people" to refer to only two persons. The descendants of Adam and Eve then lived in the northern hemisphere, the only one with land (in the southern hemisphere there is only the island of Purgatory).

But there is another hypothesis, much more interesting from an astronomical point of view. Due to the precession of the equinoxes which shifts the celestial poles, the Southern Cross was visible from Florence 7000 or 8000 years ago. Dante knew this celestial phenomenon, even if he attributed to it a longer period than the real one (36,000 years instead of 25,772, see Chapter 8). So "the first people" could be the first descendants of Adam and Eve who lived in Europe at the beginning of humanity, that is — according to what was estimated in the Middle Ages — about 7000 years ago.

It is also remarkable that the four stars of the Southern Cross are the first that the poet saw after coming out of Hell "to see the stars", feeling a bit like the "first people".

11.2 The *settentrione*

The Italian word for north derives from *septemtriones*, literally "seven oxen", because the Romans depicted the seven stars of the Big Dipper, part of the constellation Ursa Major, which are the brightest stars towards the north, like seven oxen slowly rotating to plow the sky around the north celestial pole.

At the beginning of canto XXX of the *Purgatorio* Dante created a magnificent analogy between the seven spiritual lights that guide man and the seven stars of the Big Dipper that help us orient ourselves:

> *Quando il settentrïon del primo cielo,*
> *che né occaso mai seppe né orto*
> *né d'altra nebbia che di colpa velo,*
> * e che faceva lì ciascuno accorto*
> *di suo dover, come 'l più basso face*

> *qual temon gira per venire a porto,*
> *fermo s'affisse [...]*
> (*Pur.* XXX, 1–7)

"When the Septentrion of the first heaven, which has never known setting or rising, nor the veil of any other fog than that of sin, and which made each one there aware of this duty — as the lower Septentrion does him who turns the helm so as to arrive in port —, when it came to a halt."

Just as the spiritual lights never go out, also the stars of Ursa Major are always visible (at our latitudes), because they are less than 50° from the north celestial pole and therefore always remain above the horizon. And just as the rudder of the navigator turns to take him to port, so the rudder of the Ursa Major's chariot turns around the north celestial pole.

11.3 The most beautiful stars

To give the idea of the spectacle offered by the double crown of twenty-four blessed, at the beginning of canto XIII of the *Paradiso*, Dante used an astronomical image:

> *Imagini, chi bene intender cupe*
> *quel ch'i' or vidi — e ritegna l'image,*
> *mentre ch'io dico, come ferma rupe –,*
>   *quindici stelle che 'n diverse plage*
> *lo cielo avvivan di tanto sereno*
> *che soperchia de l'aere ogne compage;*
>   *imagini quel carro a cu' il seno*
> *basta del nostro cielo e notte e giorno,*
> *sì ch'al volger del temo non vien meno;*
>   *imagini la bocca di quel corno*
> *che si comincia in punta de lo stelo*
> *a cui la prima rota va dintorno,*
>   *aver fatto di sé due segni in cielo,*
> *qual fece la figliuola di Minoi*
> *allora che sentì di morte il gelo;*
>   *e l'un ne l'altro aver li raggi suoi,*
> *e amendue girarsi per maniera*
> *che l'uno andasse al primo e l'altro al poi;*
>   *e avrà quasi l'ombra de la vera*
> *costellazione e de la doppia danza*
> *che circulava il punto dov'io era:*

poi ch'è tanto di là da nostra usanza,
quanto di là dal mover de la Chiana
si move il ciel che tutti li altri avanza.
(*Par.* XIII, 1–24)

"Let whoever wishes to grasp well what I now saw, imagine (and hold well the image, as I speak, like an immoveable rock) fifteen stars that in the various quarters of the sky shine with a clarity that overcomes every density of the air, imagine that Wain for which the first rotation turns, and that all these had made two figures in the sky like the one made by the daughter of Minos when she felt the chill of death, and let one have its rays in the other, both revolving in such a way that one begins and the other follows: and he will have almost the shadow of the true constellation and the double dance that was circling about the point where I was, for it is beyond our experience as far as the motions of the Chiana moves the heaven that surpasses all the others."

It starts with the fifteen brightest stars, which the astronomer al-Farghānī, who lived in the IX century, indicated in the 19th chapter of his work *Elements of astronomy on the celestial motions*: Sirius (α Canis Maioris), Canopus (α Argus), Rigil Kent (α Centauri), Arturus (α Bootis), Rigel (β Orionis), Betelgeuse (α Orionis), Capella (α Aurigae), Vega (α Lirae), Procyon (α Canis Minoris), Achernar (α Eridani), Aldebaran (α Tauri), Spica (α Virginis), Fomalhaut (α Piscis Australis), Regulus (α Leonis), and Denebola (β Leonis).

Today this ranking should be changed, because the luminosities of various stars have changed. In particular, Fomalhaut, Regulus and Denebola went out of the group of the fifteen brightest, while Altair (α Aquilae), Hadar (β Centauri), and Antares (α Scorpii) entered instead.

To these fifteen stars Dante added the seven of the Big Dipper, which for us always remain above the horizon. To complete the twenty-four stars that he needed for the simile with the double crown of blessed, he included in the list two other stars, α (Polaris) and β (Kochab) Ursa Minor: this last constellation was compared to a horn, in whose mouth the last two stars were found.

The reader must imagine these twenty-four stars forming two crowns in the sky, similar to the constellation of the Corona Borealis, which mythology says to be the garland worn by Ariadne, daughter of Minos, when she died abandoned by Theseus and whom Bacchus transformed into the small asterism. The two crowns are concentric and turn in opposite directions.

Even with a stretch of the imagination, the reader will have only a faint idea of what Dante saw, which is so far from our experience, the poet told us, as much as the movement of the Chiana, a slow swampy river in southern Tuscany, is far from that of the Prime Mover, that is of the heaven that rotates most rapidly (see Chapter 8).

Chapter 12

Mystery of the Milky Way

Although the Universe is very uniform on a large scale, the stars that populate it are not randomly distributed, but are gathered by the billions within dense clusters that astronomers call galaxies. There are several billions of galaxies within the cosmic horizon. Our Sun, while very important for us as it gives us light, heat, and energy (in essence it is the source of life on Earth), is just a star like any other and therefore it is part of a galaxy along with billions of other stars. Our galaxy has a disc shape on which we are dipped together with the Sun. If we look along the plane of this disc, we see many more stars than if we observe perpendicularly to it. This is the modern explanation of the luminous strip that crosses the entire sky and that we call Milky Way, or just Galaxy, with a capital letter since it is ours. Obviously the Milky Way was well known to Dante, who could easily observe it in the sky of the Middle Ages, practically devoid of light pollution produced by artificial sources.

The external galaxies, given their very large distances, are not visible to the naked eye (so Dante could not know them), with very rare but notable exceptions. In our northern hemisphere, the only one visible without the aid of an instrument is the Andromeda nebula, a galaxy similar to ours that floats in the cosmic space two and a half million light years from us. It is barely distinguishable to the naked eye as a small and faint elliptical cloud in the constellation named after the unfortunate daughter of Cepheus and Cassiopeia. While known to Arabs, Dante never mentioned it, either because he was not aware of it or because the object was not a particularly notable one. Instead, two galaxies stand out clearly in the southern sky, the Large and the Small Magellanic Clouds, both satellites of the Milky Way, respectively at 157,000 and 197,000 light years from us. Although they are dwarf galaxies, they are clearly visible to the naked eye, since they are close to us. Dante had never seen them — to observe these nebulae you should go to latitudes of less than 20°, that is for instance in Sudan, 400 kilometers south of Aswan — nevertheless he wrote of them, as we shall see later.

12.1   The Galaxy

Dante considered the problem of understanding the nature of the mysterious milky strip that "*distinta da minori e maggi/ lumi biancheggia tra' poli del mondo*" (bedecked with smaller and with greater lights, so glimmers 'tween the world's poles): the galaxy that "*fa dubbiar ben saggi*" (even the wise are led to doubt) (*Par.* XIV, 97–99). "*Per che è da sapere che di quella Galassia li filosofi hanno avute diverse oppinioni[1]*", he already noted in the *Convivio* (II, XIV, 5–8), reporting some of them, first of all a legend that he would have repeated several times in the *Commedia*:

> *Ché li Pittagorici dissero che 'l Sole alcuna fiata errò nella sua via e, passando per altre parti non convenienti allo suo fervore, arse lo luogo per lo quale passò, e rimasevi quella apparenza dell'arsura; e credo che si mossero dalla favola di Fetonte, la quale narra Ovidio nel principio del [secondo del] suo Metamorfoseos.*
> (*Convivio* II, XIV, 5–6)

"For the Pythagoreans said that the Sun at one time strayed from its path, and, passing through other regions unsuited to its burning heat, set aflame the place through which it passed, leaving there traces of that conflagration. I believe they were influenced by the fable of Phaëthon, which Ovid recounts at the beginning of the second book of the Metamorphoses."

It is the tragic myth of the son of Helios and the nymph Clymene. Phaëthon had obtained from his father to lead for a day the chariot of the Sun but, due to his lack of experience, he was not strong enough to steer the horses ("*il temo / che mal guidò Fetonte*"; *Par.* XXXI, 124–125). Having left the usual track ("*la strada / che mal non seppe carreggiar Fetòn*"; *Pur.* IV, 71–72), the chariot scorched all nature by swinging too near the Earth, to the point that Mother Earth had to implore the intervention of Zeus. In order to prevent further disasters, the king of Olympus stroke it down with a thunderbolt, together with the unwary charioteer:

---

[1] Hence it should be known that concerning this Galaxy philosophers have held different opinions.

> *quel del Sol che, sviando, fu combusto*
> *per l'orazion de la Terra devota,*
> *quando fu Giove arcanamente giusto.*
> (*Pur.* XXIX, 118–120)

"that of the Sun, which when it strayed was burned up at the prayer of the humble Earth, when Jove was mysteriously just."

But the damage had been done. The derailment of the solar chariot had left a burn on the surface of the sky, that turned into the Milky Way:

> *Maggior paura non credo che fosse*
> *quando Fetonte abbandonò li freni,*
> *per che 'l ciel, come pare ancor, si cosse;*
> (*Inf.* XVII, 106–108)

"I believe there was no greater fear when Phaëthon abandoned the reins, so that the sky was scorched, as still appears."

In the *Convivio* Dante limited himself to the above mythological quotation. In the *Commedia*, instead, he took up also another myth (but he completely neglected the most popular of the myths about the Milky Way, the one that stages Alcmena and her son Heracles, together with Hera, wife of Zeus, from whose bosom it would have come out the gush of milk responsible for the astronomical object). The story is that of the newborn Zeus whose mother Rhea tried to protect from the fury of his father Cronus. He systematically devoured his own children because of the prophecy that one of them would have dethroned him:

> *In mezzo mar siede un paese guasto,*
> *diss'elli allora, che s'appella Creta,*
> *sotto 'l cui rege fu già 'l mondo casto.*
> *Una montagna v'è che già fu lieta*
> *d'acqua e di fronde, che si chiamò Ida;*
> *or è diserta come cosa vieta.*
> *Rëa la scelse già per cuna fida*
> *del suo figliuolo, e per celarlo meglio,*
> *quando piangea, vi facea far le grida.*
> (*Inf.* XIV, 94–102)

"In the midst of the sea lies a ruined land, he said then, called Crete, under whose king the world once was chaste. A mountain is there that once was happy with water and foliage, called Ida; now it is desolate, like an outworn thing. Rhea chose it once to be the trusted cradle of her son, and, to better hide him when he cried, ordered the shouting there."

In this tale told by Virgil, the shouts were those of the Korybantes, the priests of Rhea, whom the goddess had ordered to make a great noise every time the infant Zeus, hidden in a cave on Mount Ida, began to cry, so as to prevent the father from hearing him. While the goat Amalthea fed the child, a drop of milk ended up in the sky together with the animal, one becoming the Milky Way and the other a constellation.

But Dante could not be satisfied with myths. At the same point of the *Convivio* he reported the theses of the Milesian and of the atomist philosophers, who explained the phenomenon of the Milky Way as a reflection of sunlight:

> *Altri dissero, sì come fu Anassagora e Democrito, che ciò era lume di sole ripercusso in quella parte, e queste oppinioni con ragioni dimostrative riprovaro.*
> (*Convivio* II, XIV, 6–7)

"Others, as for example Anaxagoras and Democritus, said that it was the light of the Sun reflected in that region, and they refuted the other opinions by demonstrative reasoning."

Certainly he drew on the *Doctrines of the Philosophers* of Pseudo-Plutarch, a classic in the Middle Ages, where we read:

> *[The Milky Way] is a diffuse circle, which appears continuously in the air, and due to the whiteness of its colors it is called the Galaxy, or the Milky Way. Some of the Pythagoreans say that when Phaëthon brought the world into focus, a star shooting from its place in its circular passage through the region caused inflammation. Others say that it was originally the first course of the Sun; others, that it is an image as in a mirror, caused by the Sun reflecting its rays towards the sky, and this appears between the clouds and the rainbow [...] Anaxagoras [says] that, moving the Sun under the Earth and not being able to illuminate any place, the shadow of the Earth, cast on a part of the sky, forms the Galaxy. Democritus, that is the splendor coming from the coalition of many small bodies, which, being firmly connected to each other, illuminate each other.*

Intriguing hypotheses according to Alighieri, who nevertheless sought, this time without luck, a definitive answer in Aristotle, considered one of the greatest *auctoritates* in the Middle Ages:

> *Quello che Aristotile si dicesse, non si può bene sapere, di ciò, però che la sua sentenza non si truova cotale nell'una translazione come nell'altra. E credo che fosse lo errore delli translatori: ché nella*

Nova pare dicere che ciò sia uno ragunamento di vapori sotto le stelle di quella parte, che sempre traggono quelli; e questa non pare essere ragione vera. Nella Vecchia dice che la Galassia non è altro che moltitudine di stelle fisse in quella parte, tanto picciole che distinguere di qua giù non le potemo, ma di loro apparisce quello albore, lo quale noi chiamiamo Galassia: e puote essere, ché lo cielo in quella parte è più spesso, e però ritiene e ripresenta quello lume. E questa oppinione pare avere, con Aristotile, Avicenna e Tolomeo.

(*Convivio* II, XIV, 7)

"What Aristotle said on this matter cannot be known with certainty because his opinion is not the same in one translation as in another. I believe that this is due to an error on the part of the translators; for in the *New Translation* he seems to say that it is a collection of vapors beneath the stars in that region, which attracts them continuously; this does not seem to have any foundation in truth. In the *Old Translation* he says that the Galaxy is nothing but a multitude of fixed stars in that region, so small that we are unable to distinguish them from here below, though from them originates the appearance of that brightness which we call the Galaxy; this may be so, for the heaven in that region is denser, and therefore retains and throws back this light. Avicenna and Ptolemy seem to share this opinion with Aristotle."

Therefore, the poet favored the hypothesis of the grouping of stars, later confirmed by Galileo Galilei's observations of 1609, published in the *Sidereus Nuncius*, a work in Latin with which the Pisan scientist disclosed his astonishing astronomical observations:

Quello che [. . .] osservammo è l'essenza o materia della Via Lattea, la quale attraverso il cannocchiale si può vedere in modo così palmare che tutte le discussioni, per tanti secoli cruccio dei filosofi, si dissipano con la certezza della sensata esperienza, e noi siamo liberati da sterili dispute. La Galassia infatti non è altro che un ammasso di innumerabili stelle disseminate a mucchi.

"What was observed by us [. . .] is the nature or matter of the Milky Way itself, which, with the aid of the spyglass, may be observed so well that all the disputes that for so many centuries have vexed philosophers are destroyed by visible certainty, and we are liberated from wordy arguments. For the Galaxy is nothing else than a congeries of innumerable stars distributed in clusters."

As for the sidereal Universe, Dante, who had drunk himself at the Aristotelian–Thomist wisdom to the point of preferring it, on an astronomical level, to that of Ptolemy, perhaps because the latter appeared to him not very philosophical and too technical, could not have gone beyond, not even exercising his extraordinary imagination and creativity. To consolidate the knowledge on the shape and composition of the Milky Way and at the same time ascertain its role as a member of the populous family of the nebulae, it would have been necessary to wait until the XVIII century. First of all, the Copernican concept of a humanity no longer placed at the center of the Universe, as an effect of its descent from God, had to be metabolized; and it was not easy, given the theological implications. Furthermore, a sufficiently powerful instrument was needed to allow observation of the deep sky, where Galilei's *perspicillum* stayed blind.

As we have already mentioned (see Section 4.4) Immanuel Kant, drawing inspiration from an intuition of the English thinker Thomas Wright, was the first to propose an infinite space populated by islands of matter, which we now call galaxies. These *"island universes"* — an expression used by the German naturalist and geographer Alexander von Humboldt to popularize Kant's model — were observed repeatedly, along with the Milky Way, by the Anglo-German astronomer Friedrich Wilhelm Herschel and his sister Caroline.

The result was the creation of a scientific case rather than the solution of a problem. The question concerned the nature of the cosmos as a whole: was it a single immense distribution of stars lost in infinite space, with a central concentration in correspondence with the Sun, with all respect due to Copernicus, or instead an archipelago of galaxies? To solve the dilemma, an instrument was needed capable of measuring endless distances, completely inconceivable at the time of Dante, when the Universe enclosed by the Fixed Stars had a radius slightly greater than the orbit of Saturn (let us say, in modern terms, one light-hour at maximun compared to 50 billion light-years of the cosmological horizon estimated today).

This magical instrument, represented by the stars with hiccups (the Cepheids, a type of pulsating stars whose brightness varies periodically with great regularity), was identified at the beginning of the XX century by Henrietta Leavitt and masterfully used by Edwin Hubble. In 1924 the American astronomer measured the distance of some extragalactic Cepheids, and therefore of the stellar systems containing them, finally clarifying what galaxies are and how the luminous matter is clustered in celestial spaces. An idea of the cosmos very far from that of Dante and, we must add, even from the one in force today. We have in fact ascertained, in the last fifty years, that the matter interacting with light, the one which "*il Sole e l'altre stelle*", "the Sun and the other stars" are made of (according to a vision that — it is to be believed — would have greatly disturbed the *Weltanschauung* in Dante's theological key), constitutes only a small slice of the cosmic cake of matter–energy, where dark ingredients prevail: still mysterious entities unable to interact with light.

One wonders how the great poet would have treated these new ingredients in the context of the metaphors of which light is a positive and divine manifestation. But Dante, closed in his Aristotelian cosmos, could not even imagine it. There is nothing wrong with it. Science is a form of progressive knowledge, which takes advantage of previous work to achieve better results and does not know the terms truth and perfection (in the etymological sense of something that need not be changed). The new science born with Galilei could not belong to a medieval spirit (albeit a very high one), striving to know more "*how to go to heaven*", in the words of the Pisan genius in the letter sent in 1615 to Christina of Lorraine, grand duchess of Tuscany, than "*how heaven goes*".[2]

12.2 The Magellanic Clouds

As we have already said, Dante perhaps spoke of the Magellanic Clouds, the two small satellite galaxies of the Milky Way, when in canto VIII of *Purgatory*, after listening to Nino Visconti, he turned his gaze to the southern pole:

> *Li occhi miei ghiotti andavan pur al cielo,*
> *pur là dove le stelle son più tarde,*
> *sì come rota più presso a lo stelo.*
> *E 'l duca mio: "Figliuol, che là sù guarde?"*
> *E io a lui: "A quelle tre facelle*
> *di che 'l polo di qua tutto quanto arde".*
> *Ond'elli a me: "Le quattro chiare stelle*
> *che vedevi staman, son di là basse,*
> *e queste son salite ov'eran quelle".*
> (*Pur.* VIII, 85–93)

"My greedy eyes were still seeking the sky, especially where the stars are slowest, like a wheel closest to the axle. And my leader said to me: 'Son,

---

[2] "*Come si vadia al cielo, e non come vadia il cielo.*"

what are you staring up there?' And I to him: 'at those three little torches with which the pole here is all aflame.' Then he to me: 'The four bright stars that you saw this morning are low over there and these have gone up where those were.'"

Apart from the unanimous assimilation with the three theological virtues (faith, hope, and charity), the astronomical identification of the "*facelle*" is more difficult. However Dante said that they were circumpolar and at the same declination, but with the opposite right ascension (therefore at 180°) of the four stars he had seen in the morning, the Southern Cross (see Section 11.1). As hypothesized by the astronomer Ernesto Capocci, who for a long time directed the Astronomical Observatory of Capodimonte in Naples in the period that saw the transition from the Bourbon to the Savoy kingdom, it could be Achernar, the brightest star in the constellation of Eridanus, and the two Magellan Clouds, of which Dante might have learned from the Arabic celestial globes, in which they were indicated as White Ox (Al Bakr in Arab).

We have already spoken of the well known fact that galaxies are made of stars; information that, however, was certainly not obvious in Dante's time, in which the sky could only be observed with the naked eye. It is therefore worth remembering what the Vicentine navigator Antonio Pigafetta wrote about the Clouds in his report on the first voyage around the world, where he recounted the enterprise of Ferdinand Magellan, which he participated from 1519 to 1522:

> *The Antarctic pole is not as starry as the Arctic one. You see many small stars, congregated together, they make like two nebulae that are not very separated from each other and slightly blurred.*

So Pigafetta, even if he had observed them with the naked eye, clearly said that the two objects, later improperly baptized Magellanic Clouds in honor of his commander, were composed of numerous stars, faint because of the distance and gathered together.

Ninety years would still have to pass before Galilei in 1609 pointed a telescope towards the sky and became the first to give definitive confirmation that the Galaxy is made of stars.

Chapter 13

An unlimited but finite Universe

We discuss here a very exciting possibility from a scientific point of view: Dante had imagined a curved Universe, unlimited but finite, which may anticipate an Einstein hypersphere, or 3-sphere. Here the matter becomes much more complicated and intriguing, bearing witness of Dante's scientific insights (not to be confused with any formal knowledge, however, as some scholars have forcibly maintained). These insights go far beyond the state of the art of his time and demonstrate, where necessary, as the mind of a poet can give birth to illuminating ideas transcending current understanding.

The topic is objectively complex and somewhat controversial. Dante himself said so in the opening words of canto II of Paradiso. Paraphrasing it, the warning sounds more or less like this: if you are not equipped to follow me into the deep sea of knowledge, go home and forget it:

> *O voi che siete in piccioletta barca,*
> *desiderosi d'ascoltar, seguiti*
> *dietro al mio legno che cantando varca,*
> *tornate a riveder li vostri liti:*
> *non vi mettete in pelago, ché forse,*
> *perdendo me, rimarreste smarriti.*
> *L'acqua ch'io prendo già mai non si corse;*
> *Minerva spira, e conducemi Appollo,*
> *e nove Muse mi dimostran l'Orse.*
> (*Par.* II, 1–9)

"O you who in little barks, desirous of listening, have followed after my ship that sails onward singing: turn back to see your shores again, do not put out on the deep sea, for perhaps, losing me, you will be lost; the water that I enter have never before been crossed; Minerva inspires and Apollo leads me, and nine muses point out to me the Bears."

So let us start step by step. Unlike Hell and Purgatory, which are physically connected with the Earth and therefore with that subsystem of the Aristotelian cosmos where time reigns with its failures, Paradise is found in the eternal and immutable skies. An ethereal world, separated from the physical one by a sphere of fire and built, in its first part, in the image of Ptolemy's geocentric system: seven spheres of increasing radius centered on the Earth, each containing the previous one and characterized by a wandering star (Moon, Mercury, Venus, Sun, Mars, Jupiter and Saturn, according to an order that respects the apparent motions observed by astronomers). They rotate with increasing speed and are composed of a substance called ether, introduced by Aristotle in antithesis to the four elements of Empedocles that give life to terrestrial structures, corruptible and imperfect (see Chapter 4). As they move they shine, emit soft sounds, and influence men and their events (according to a vision rooted in classical judicial astrology).

These seven spheres are followed by the spheres of the Fixed Stars and of the First Mover, a sort of engine for the complex celestial clock that Dante, not having the obligatory confirmation of appearances, simplified by limiting it to pure circular motions. It is capable of transmitting movement, understood as a growing aspiration to the highest virtue. In total, therefore, nine spheres: the last, whose invention Dante erroneously attributed to Aristotle, was actually introduced by medieval philosophers to insert a principle of cause and effect within a cosmological framework permeated with theological concepts. It is not the only erroneous attribution of the poet, further proof that not all of his vast knowledge was based on primary sources. For example, Dante attributed the discovery of precession and ideas about the nature of the Milky Way to Ptolemy and not to Hipparchus of Nicaea, who lived four centuries earlier.

Incidentally, it is worth emphasizing how in 1320 Dante revised his poetic-theological explanation of the formation of Hell in a physical key. The poet, who had then completed his monumental work, was a guest of Cangrande della Scala in Verona. Scientific questions were debated at the court of this young Ghibelline prince, according to a model already experimented by Frederick of Swabia. On a cold January morning, Dante entertained Cangrande together with a large group of canons and lay people with his theory on why the emerged lands were always higher than the level of the oceans. According to Aristotelian theory, in fact, the sphere of water is external to the sphere of the earth. It is therefore a question of either denying the evidence, arguing that water completely covers the earth, or giving an account of the existence of the emerged lands, appealing to a lifting action of earthly masses by the planetary spheres: a sort of tide that actually exists, although it is completely without consequences (the forces that have shaped the planet's crust are mainly endogenous or at the most the result of catastrophic and occasional impacts), and that establishes some (heretical) physical connection between the two disjointed spheres of Aristotelian cosmology, Earth and sky.

The Latin text of this *disputatio* would have been published in Venice with the title of *Questio de aqua et terra* only two centuries later, proving the embarrassment aroused among the poet's exegetes for the significant change from the cosmogonic theses exposed in the *Commedia*, but also for the rationalistic attitude which, in its application to natural philosophy, favored the inductive over the deductive method typical of mathematics. Dante then, relying on the authority of the *Holy Books*, took a cautious attitude regarding the search for truth, because not everything can be understood, he argued, by the human mind. Without the help of Beatrice, who participated in divine wisdom, human reason could not venture into the mysteries of the Universe.

Let us go back to the First Mover, of which Dante described the peculiar properties through the words of Beatrice:

> *e questo cielo non ha altro dove*
> *la mente divina, in che s'accende*
> *l'amor che 'l volge e la virtù ch'ei piove.*
> *Luce e amor d'un cerchio lui comprende,*
> *sì come questo li altri; e quel precinto*
> *colui che 'l cinge solamente intende.*
> (*Par.* XXVII, 109–114)

"and this heaven has no other where than the mind of God, in which is kindled the love that turns it and the power that it rains down. Light and love enclose it with one sphere, as this does all the others, and that girding only he who girds it understand."

Something singular happened there. Dante told it in the following canto XXVIII:

> *un punto vidi che raggiava lume*
> *acuto sì, che 'l viso ch'elli affoca*
> *chiuder conviensi per lo forte acume;*
> *e quale stella par quinci più poca,*
> *parrebbe luna, locata con esso*
> *come stella con stella si collòca.*
> *Forse cotanto quanto pare appresso*

alo cigner la luce che 'l dipigne
quando 'l vapor che 'l porta più è spesso,
 distante intorno al punto un cerchio d'igne
si girava sì ratto, ch'avria vinto
quel moto che più tosto il mondo cigne;
 e questo era d'un altro circumcinto,
e quel dal terzo, e 'l terzo poi dal quarto,
dal quinto il quarto, e poi dal sesto il quinto.
 Sopra seguiva il settimo sì sparto
già di larghezza, che 'l messo di Iuno
intero a contenerlo sarebbe arto.
 Così l'ottavo e 'l nono; e chiascheduno
più tardo si movea, secondo ch'era
in numero distante più da l'uno;
 (*Par.* XXVIII, 16–36)

"I saw a point that was radiating light so sharp that any eye into which it shines must close at its piercing intensity, and whatever star from here seems smallest, would seem like the Moon if placed near it, as in the sky star is placed next to a star. Perhaps as closely as the halo seems to gird the light that projects it, when the vapor that carries it is thickest, so around the point as circle of fire was turning so swiftly that it would have surpassed the motion that girds the world most swiftly, and this was girt about by another, and that by a third, and the third then by the fourth, by the fifth the fourth, and then by the sixth the fifth. Beyond there followed the seventh, already so expanded in circumference, that the messenger of Juno would be too narrow to contain it. Thus the eighth and the ninth, and each moved more slowly as it was numerically more distant from the one."

The poet saw a light reflected in Beatrice's eyes and turned to look at the light source like someone who had seen a torch reflected in a mirror and moved his gaze on the real object. A light hurt him, very intense and at the same time tiny: a dimensionless point (a geometric entity already known to Euclid), so much so that the faintest of the stars would appear larger than the Moon in comparison. A flaming diffuse halo revolved around it, with a speed that exceeded the already remarkable one of the First Mover. It was made up of nine circles, of decreasing radius as the speed of rotation increased (therefore, an inverse situation to that of the planetary spheres). These circles were actually so many spheres for the poet, as he had already clarified in a note from the *Convivio*.

What does all that mean? Dante struggled to understand it, but Beatrice came to his aid by offering him the key to the reversal of behavior between the planetary spheres and the choirs of light that surrounded the central singularity. The velocity of motion of these angelic spheres (which had a different order in the *Commedia* compared to that adopted in the *Convivio*, as proof of a constant reworking of ideas) grew closer to God, because the perception of his love increased, so that the innermost, formed by the Seraphs, was the fastest. The poet was perplexed (and so are we) because this explanation collided with the perception of the planetary spheres that had the Earth as their center. But Beatrice insisted: the cause was virtue and her search for divine love. The extension did not count; the Empyrean was enclosed in the First Mover, the one that contained the most love and wisdom, whose fire was God:

> *Per che, se tu alla virtù circonde*
> *la tua misura, non all'apparenza*
> *delle sustanze che t'appaion tonde,*
> *   tu vederai mirabil consequenza*
> *di maggio a più e di minore a meno,*
> *in ciascun cielo, a süa intelligenza.*
> (*Par.* XXVIII, 73–78)

"Thus if you circumscribe your measure with the power, not the apparent size, of the circling substances that here appear to you, you will see a marvelous correspondence of greater to more and of smaller to less, between each heaven with its Intelligence."

Then for Dante everything was clear, as he tells us with one of his elegant similes:

> *Come rimane splendido e sereno*
> *l'emisperio de l'aere, quando soffia*
> *Borea da quella guancia ond'è più leno,*
> *   per che si purga e risolve la roffia*
> *che pria turbava, sì che 'l ciel ne ride*
> *con le bellezze d'ogne sua paroffia;*
> *   così fec'ïo, poi che mi provide*
> *la donna mia del suo risponder chiaro,*
> *e come stella in cielo il ver si vide.*
> (*Par.* XXVIII, 79–87)

"As the hemisphere of air is left brilliant and clear when Boreas blows from his milder cheek, so that he purges and dissolves the dregs that previously clouded it, and the sky laughs to us with its beauties from its very parish:

so I became when my lady provided me with her clear replay and like a star in the sky the truth became visible."

But it is not clear to us, unless we resort to a geometric model that contemplates the concept of hypersphere, completely alien to mathematics and geometry coeval with Dante. Let us see why.

The First Mover is the main circle that surrounds all the planetary spheres, which in turn embrace the Earth centered on Lucifer, and the Empyrean, where the angelic choirs gradually close in nine circles, as many as the heavens, around a point coincident with God. Two antithetical foci, therefore, are embraced by the First Mover, from which both can be seen with all the structures associated with them. The unifying trait is the velocity of rotation, which grows monotonously from the Moon to the Seraphs: a sort of coordinate that measures an implied property, the different reception of divine love, and a sort of cosmic law that Dante anticipated in the *incipit* of *Paradise*:

> La gloria di colui che tutto move
> per l'universo penetra, e risplende
> in una parte più e meno altrove.
> (*Par.* I, 1–3)

"The glory of Him who moves all things penetrates through the Universe and shines forth in one place more and less elsewhere."

How can all this be understood? Before trying to solve the dilemma by resorting to a solution invoked just a century ago, objectively difficult to digest and not yet shared by the whole community of Alighieri's scholars, we must note that, while in Hell and Purgatory the damned and the souls have their own fixed location, so that Dante could easily establish and contextualize the connection between faults and pains, in Paradise the blessed are free to move in order to enjoy any place. God does not place restrictions on them, the poet let us know. However, for the entire duration of the journey in which Beatrice guided him, everyone settled himself in the place of his competence. It was certainly a solution adopted by Dante to make the narrative coherent. But it can also be read as a way to fix the scale of the coordinate: rotation velocity = density of divine virtue, which marks the distance from Lucifer on one side and from God on the other.

One way to rationally interpret this set of statements by Dante is to admit that he had intuited the properties of the hypersphere and translated them into a model, projecting it into the three dimensions accessible to be represented and told. What is a hypersphere? In our case it is a three-dimensional sphere, in a four-dimensional space, and it is the locus of the points that have all the same distance (which we will call radius R) from a given point O, called the center of the hypersphere. Nothing strange. By scaling to the immediately lower dimension we find that the equivalent geometric figure is the simple spherical surface, and then the circumference going down again until the space is reduced to a plane. Whoever is confined to living within one of these figures will experience a Universe of finite volume (for example, the volume of a sphere and, if the creature has only two dimensions, the surface of a sphere) but unlimited.

The creatures who are bound to live on the frontier (flat if it is a spherical surface and one-dimensional if a circumference) will have the following experience. Their world too has a finite volume; however it appears unlimited. The toy train that rattles on its ring track is not obliged to stop after one turn. It is understood that the spaces represented by the two borders of our examples (spherical surface and circumference) have a non-zero (Gaussian) curvature. They are not Euclidean like the volumes they enclose: in order to map them, their possible inhabitants would have to resort to coordinate systems other than the classic Cartesian coordinates. But this does not scare a modern mathematician or even Ptolemy, who had prefaced the strictly astronomical section of his *Almagest* with a complete treatise on spherical trigonometry.

We thus come to the hypersphere. We are unable to represent it in two dimensions, or in three, using the tricks of perspective or false colors to visualize it. But not all is lost. We can figure out some of its properties using the method described by Edwin Abbott Abbott in his brilliant work of 1884 entitled *Flatland*: a fantastic tale about multiple dimensions.

The idea of the British writer, theologian, and pedagogue, is to explain the perception of the world in $n + 1$ dimensions by people who live in a subset of n dimensions. Abbott imagined a flat society of individuals confined on a floating plane in three-dimensional space, where their world met a sphere. Cut by the plane, the solid left an imprint that started with a point and continued with circumferences of increasing radius up to a maximum, and then returned to decrease until it became again a point before it disappeared. So, what does a hypersphere do when it encounters our three-dimensional space? It leaves as footprints a sequence of spherical surfaces that start from a point, reach a maximum size, and return to become a point. This happens because the hypersphere has to sacrifice a coordinate (or a combination of them). Basically, we can represent it through two hemispheres of equal radius and tangent to each other, each of which incorporates ever smaller circles until it is reduced to a point.

Well, this representation is extraordinarily similar to Dante's description of Paradise, where the coordinate that labels the individual circles is precisely the growing light of God. The circles are somewhat quantized, nine for each part including the Prime Mover, which is common to the two hemispheres and from which the poet can always see in both directions in the same way, no matter where he is. The two central points are symmetrically occupied by the angel of evil (Lucifer) and by God, and each is both focus and edge of the world: a Universe that appears finite and yet at the same time unlimited. In this way, the difficulties encountered by medieval theologians in accepting the Aristotelian vision of a limited physical world are resolved: its existence assumes a container which, as unlimited and timeless, creates an unacceptable competition with God himself. Thus, as the Jesuit Father José Gabriel Funes wrote, "*the divine mystery will be both impregnable sphere and concentric center of reality, containing and contained, transcendent and immanent, as a dazzling aphorism in the medieval Book of the twenty-four Philosophers said: 'God is an infinite sphere, whose center is everywhere and the circumference nowhere.'* "

It is difficult to explain how Dante could have thought of such a daring geometric solution five centuries before great mathematicians such as Carl Friedrich Gauss and Bernhard Riemann, and then utilized them in his cosmology even anticipating Einstein in the use of four coordinates: three for space and one more, the divine virtue, which for the German physicist will be instead time; a *sui generis* coordinate as it is bound to increase only (arrow of time).

An intuition, albeit not supported by any formal elaboration? It is hard to affirm this, given the condition of relative backwardness of mathematical speculation in the XIV century which, although illuminated by some flashes (such as the succession of the Pisan Leonardo Fibonacci, who lived between the XII and XIII centuries), still remains in the shadow of Euclid and his Arab followers. Denying it, however, would expose us to the need of finding another way, for now unknown, to account for Dante's fantastic construction. Perhaps the solution is an inextricable intertwining of fantasy, poetic genius, and lucky coincidences. Or perhaps the Anglo-American poet Thomas S. Eliot, Nobel Prize for Literature in 1948, was right when he said: "*it is not essential that Dante's almost unintelligible astronomy be understood.*"

An idea to grasp how Dante could have thought of a curved universe can come from what Brunetto Latini wrote in his encyclopedic work *Li livres dou tresor*, composed around 1265, in which, to describe the sphericity of the Earth, he imagined two riders starting from one same place at the same time and moving with identical speed in different directions. If there were no seas and mountains — he observed — they would reach the very same point, at the antipodes. So, the spherical surface of the Earth is described standing on it, not from the outside, and this is what happens. If they start from the north pole, they would ride south along different sides of the same meridian, cross increasingly larger parallels to the equator, then in the southern hemisphere increasingly smaller parallels and meet again at the south pole. Similarly Dante started from the center of the Earth and crossed ever larger celestial spheres up to the First Mobile, from which he saw the Empyrean with ever smaller angelic circles up to the point where God was. This vision calls to the mind of an expert in general relativity the curved, unlimited, but finite universe, predicted by Einstein.

Dealing with possible advances of physical concepts of relativity by Dante we should mention another one, concerning the relativity of reference frames, which was later proposed by Galileo Galilei. It must be said that, apart from the two lectures about the figure, site, and size of Dante's Hell (see Section 6.1), there are very few references to Dante in the rest of the work of Galilei. Instead, there is a suggestion of a strong link between the two great Italians regarding one of the most important discoveries of the Pisan physicist, the principle of relativity. Galilei exposed it in the *Dialogo sopra i due massimi sistemi*, published in 1632, using the classic example of the ship. A passenger below deck will not notice that he is traveling if the ship (*"gran navilio"*) moves straight ahead at a constant speed (obviously in a flat sea). Of this principle, fundamental for the work of Newton and then of Einstein, there is perhaps an anticipation in canto XVII of the *Inferno*. Having to go down from the seventh to the eighth circle, Dante and Virgil used a flying monster, Geryon, on whose back they both climbed. It will be an experience that requires courage, was the warning of Virgil; and in fact Dante showed how much he was scared by citing the tragic examples of the flights of Phaëthon and Icarus. But then, when he found himself in the air without any reference, although in terror he remarked that he did not realize he was flying, if not for the wind that hit his face from underneath.

> [...] *quando vidi ch'i' era*
> *ne l'aere d'ogne parte, e vidi spenta*
> *ogne veduta fuor che de la fera*
>   *Ella sen va notando lenta lenta;*
> *rota e discende, ma non me n'accorgo*
> *se non che al viso e di sotto mi venta.*
>   (*Inf.* XVII, 112–117).

"when I saw that I was in the air on every side, and every sight put out save that of the beast. It goes along swimming slowly, slowly; it wheels and descends, but I perceive its motion only by the wind on my face from below."

Exactly the same experience as that of the traveler in the Galilean ship. Another intuition that, due to its extraordinary nature, leaves us amazed at Dante's ability to read the physical world.

Chapter 14

Parallel Universes

In the *Vita Nova*, in the *Convivio*, and in the *Commedia* , Dante dealt with topics of astronomical nature, but he never touched on the question of the plurality of worlds, that is whether there are other forms of life and intelligence in the cosmos. Even the complex architecture of Paradise, with its daring geometries and its angelic hosts, is an integral part of a single world surrounded by the firmament, where it is both contained and container in the grace of divine majesty.

Although he knew of Epicurus, as it can be inferred from the criticism made on him in canto X of *Inferno* (vv. 14–15) concerning the alleged denial of the immortality of the soul, Dante did not even mention the hypothesis of a Universe consisting of many worlds. Starting from ideas developed in the previous two centuries, the Samo philosopher, who lived between the IV and III centuries BC, had developed a theory full of suggestion, with some aspects of extraordinary modernity, which we can summarize as follows: 1. matter is composed of indivisible entities (atoms); 2. nature is subject to a process of continuous change due to the aggregation and disintegration of atoms; 3. there is no creative intelligence; 4. the Universe is eternal and infinite. So he wrote:

> *There are infinite worlds, both similar and different from our world [. . .] and we must believe that in all worlds there are creatures, plants and all the other things that we see here in our world.*
> (Epicurus, *Epistle to Herodotus*)

Aliens *ante litteram*? Obviously not! Rather inhabitants of what today we might call parallel Universes. Epicurus, in his multiplying worlds, imagines them as replicas (invisible and unknowable) of a complete system like ours, in which, according to Ptolemy's synthesis, the sphere of the Fixed

Stars is an essential ingredient: a sort of container on top of which the First Mover is placed.

Plato and especially Aristotle were opposed to Epicureanism and rejected the thesis of the infinite worlds with philosophical arguments. Aristotle, for example, noted that the existence of distinct worlds would generate uncertainty in the motion of the elements towards their natural places: a stone is mainly made up of earth and therefore tends to fall downwards, but if there were many Earths the direction of natural motion would be confused. As we know, two millennia later Newton would have solved brilliantly the problem. The major objections, however, were of a religious nature. Plato reasoned that the uniqueness of the creator, which he postulated, implied the exclusivity of the creative act. Aristotle, on the other hand, rejected the multiplicity of First Movers, resulting from the hypothesis of the plurality of worlds. A difficulty that Epicurus had to overcome in one leap, denying the existence of God and entrusting the task of continuous creation to atoms, their laws, and chance.

Even the Pythagoreans had cultivated the idea of an extraterrestrial life, without however invoking new worlds, limiting it to the Moon, the classic limit between the terrestrial world, corruptible and imperfect, and the celestial one, complete and immutable. In fact, according to a suggestion that would have various interpretations in the centuries to come, including the XX century, the Moon has often been considered a habitat for extravagant life forms. Dante himself made a *sui generis* use of the Moon (*Par.* II), justified by the fact that its heaven borders the two worlds; therefore this celestial body could tolerate changes and imperfections.

Despite the disinterest shown by the great poet for a hypothesis that would have undermined the entire castle built with the *Commedia* , the debate on the plurality of worlds could not fail to interest Christian writers and thinkers, due to the close relationship with the themes of creation and of revelation (as set out in the *Old Testament*) and with redemption (in the *New Testament*). The question was far from simple and without pitfalls, because it was a matter of reconciling a single creative act defined in time with the omnipotence and eternity of God. The opponents of the plurality of worlds, such as Thomas Aquinas, who had prevailed during the early Middle Ages, suffered a first defeat at the end of the XIII century. About forty years before Dante, then in exile, put his hand to the *Commedia* , Aristotle's natural philosophy was banned, as we already mentioned; in particular the propositions guilty of limiting divine power, including the one that reads: the First Cause cannot create many worlds.

But criticizing Aristotle's conclusions was not enough. His arguments had to be demolished. Among others, the French scientist, philosopher, and theologian, Nicole Oresme, took care of it, arguing:

> God can and could in His omnipotence make another world besides this one or many others similar or even different. There is no Aristotle or anything else that can completely prove the opposite.
> (*Book of Heaven and the World* I, 24)

But then he took it all back, ruling: "*There has never been nor will there ever be more than a single corporeal world*". In other words, God can do everything, but for inscrutable reasons he has chosen to create one and only one world, as we seem to understand from the Gospel of John in the *Scriptures*. The human mind pawed, but a rigid interpretation of the sense of loyalty to God tied the hands to science. Of this debate that pervaded the philosophical schools of Northern Europe there is no mention in Dante; he, however, could not fail to know it for the breadth of his readings and for the richness of his contacts. The poet was in this sense strictly orthodox and to some extent conservative.

In the XV century, the German humanist Nicholas of Cusa widened the problem of plurality, confined by the thinkers of Humanism to the inanimate worlds only, extending it to the more delicate question of the inhabitants of these worlds, putting also the planets into play. About a century later, in 1584 Giordano Bruno resumed this thesis, arguing in the third dialogue of *De l'infinito, universo e mondi*:

> *Uno dunque è il cielo, il spacio immenso, il seno, il continente universale, l'eterea regione per la quale il tutto discorre e si muove. Ivi innumerabili stelle, astri, globi, soli e terre sensibilmente si veggono, ed infiniti ragionevolmente si argumentano. L'universo immenso ed infinito è il composto che resulta da tal spacio e tanti compresi corpi.*

"One is therefore the sky, the immense space, the bosom, the universal continent, the ethereal region through which everything flows and moves. There innumerable stars, celestial bodies, globes, Suns and Earths are seen thanks to the senses, and reasonably considered infinite. The immense and infinite Universe is the compound that results from this space and from the many bodies that are included in it."

For his ideas, of extraordinary strength, Bruno was condemned as a heretic and led to the stake in Campo de' Fiori on February 17, 1600, with his mouth tightened by biting. An unjust and cruel punishment for a crime of opinion which, however, had a disruptive effect in a lacerated and then divided Christian world. In fact, the pluralistic cosmology vigorously supported by Bruno presented many problems for theology, as a Breton Franciscan of the mid-XV century had already explained:

> As for the question of whether Christ by dying on this Earth could redeem the inhabitants of other worlds, I reply that he is capable of doing so even if the worlds were infinite, but it would not be adequate for him to go to another world where he must die again.

In 1543, the *De revolutionibus orbium coelestium* by Nicolaus Copernicus was published: it was the first step of a long road towards the progressive liberation of scientific investigation from religious interferences, a path that can be said to be completed when, according to a famous anecdote, Pierre Simon de Laplace, in a reply to Napoleon's question as to why he had never mentioned God in his *Exposition du système du monde* of 1796 the great scientist, the great scientist replied that he had no need of this *"gracious hypothesis"*.

It is interesting to note that none of the outstanding figures who helped ferry the Copernican revolution to Newton was a true supporter of the plurality of worlds; or, at least, not in a modern sense. Galilei, who during his life rarely distinguished himself for prudence, in the face of this question exhibited an unusual measure. In his work *Around the sunspots* he wrote:

> Se poi si possa probabilmente stimare, nella Luna o in altro pianeta esser viventi e vegetabili diversi non solo da i terrestri, ma lontanissimi da ogni nostra immaginazione, io per me nè lo affermerò né lo negherò, ma lascerò che più di me sapienti determinino sopra ciò, e seguiterò le loro determinazioni.
>
> (*Le opere di Galileo Galilei*, National Edition)

"If we can probably estimate that, in the Moon or in another planet, there are living and vegetable beings different not only from the terrestrial ones, but very far from all of our imaginations, I for myself will neither affirm nor deny it, but I will let more than me knowledgeable people determine above this, and I will follow their judgements."

A caution that clashed when compared with the courage of Tommaso Campanella. Praising Galilei from the Neapolitan prison of Castel Nuovo where he was held in chains, he ventured a daring maneuver to reconcile the needs of faith with the pluralist hypothesis.

Meanwhile, the new science born of Galilei was finding its prophet in Isaac Newton and a fertile ground in the kingdoms of Northern Europe.

Even the question of the plurality of worlds would have had to deal with this new cultural and methodological context and with a world which, if it did not free the subject from the vexed question of divine attributes and from the hesitations of religious faith, at least relieved it from the fear of the Inquisition. For four centuries, scientists and philosophers, besides dreamers, would have speculated on the existence of life in the Solar System, with a real peak of interest after the discovery of the alleged channels of Mars by Giovanni Virginio Schiaparelli, in the second half of the XIX century.

Thus, in the wake of the Cartesian Bernard Le Bovier de Fontenelle and on the ruins of Aristotelian dualism, supplanted by a unitary vision of the cosmos, the Brunian conception of the infinite worlds will reappear: a faith that will have followers of great fame — such as Kant, Laplace, Herschel, Nicolas Camille Flammarion, and Percival Lowell — and which today arouses the growing interest of astronomers, physicists, and biologists.

Chapter 15

Astrology

Under the heading "astrology", the *Treccani* vocabulary reads as follows:

> *Divinatory art, once considered a science, which presumes to determine the various influences of the stars on the earthly world and on the basis of them predict future events or give an explanation of past events that remain unknown.*

Immediately afterwards it clarifies that, up to the XVII century, the term was used as a possible synonym for astronomy, hence

> *[. . .] the use of expressions: judicial astrology, the art of reading the judgment of heaven on earthly events; spherical astrology, the science that studied the course of the stars without taking into account their influences.*

It might seem clear, but the story is not really that simple. In the West, predictive astrology has very ancient roots and a development that is not at all linear. Its classical form was canonized in its rules (totally groundless in the sense of Galilean science, but formally rigorous from a mathematical point of view) by Claudius Ptolemy, who expounded them in the *Tetrabiblos*, another great work of the Alexandrian astronomer. Actually, in the early medieval Christian world astrology had lost credit due to the obvious conflict, reported among others by Augustine, the influential bishop of Hippo, between its determinism and the freedom of believers to choose their actions, which translates into moral responsibility. In this regard Dante, in canto XVI of the *Purgatorio*, made the wise and valiant Marco Lombardo say:

> *Voi che vivete ogne cagion recate*
> *pur suso al cielo, pur come se tutto*

> *movesse seco di necessitate.*
>   *Se così fosse, in voi fora distrutto*
> *libero arbitrio, e non fora giustizia*
> *per ben letizia, e per male aver lutto.*
>   *Lo cielo i vostri movimenti inizia;*
> *non dico tutti, ma, posto ch'i' 'l dica,*
> *lume v'è dato a bene e a malizia,*
>   *e libero voler [. . .]*
>   *(Pur. XVI, 67–76)*

"You who are alive still refer every cause up to the heavens, just as if they moved everything with them by necessity. If that were so, free choice would be destroyed in you, and it would not be justice to have joy for good and mourning for evil. The heavens begin your motions; I do not say all of them, but, supposing I say it, a light is given you to know good and evil, and free will [...]"

In the second millennium, however, with the repêchage of Aristotelian cosmology and its integration into Christian theology, astrology found full legitimacy in a scheme in which the celestial intelligences and the spheres they animated became the interface between God (the *"first cause"* for the thinker of Stagira) and the corruptible and imperfect world encircled by the sky of the Moon. No longer a superstitious reading of wandering bodies on the plot of the firmament, but rather a para-religious doctrine which nevertheless preserved the unscientific codes and languages of the ancient tradition. With this renewed purpose it was welcomed by Albert the Great and his disciple Thomas Aquinas, two of the fundamental references for the formation of Dante's thought.

At the time the poet lived, no man of culture in the Christian West (educated in the faith and in the disciplines of the *trivium* and *quadriv-ium*[1]) could doubt the existence of astral influences attributable to God, capable of moving the skies and affecting human affairs. On the other hand, the rejection of predictive astrology (apparently) based on reason and not aimed at recognizing God in things was clear (the criticism did not concern the system of rules, in fact completely arbitrary, but the epistemological premises). Therefore, we are not surprised by the lexical and substantial mixture between astronomy and astrology that we find in Dante's works.

---

[1] Division of the seven liberal arts, grouped in grammar, logic, and rhetoric (trivium), and in arithmetic, geometry, music, and astronomy (quadrivium). They are based on thinking skills, at variance with practical arts such as architecture and medicine.

Before going into some detail, it is worth remembering that during Renaissance astrology underwent a new drift towards an increasingly marked superstitious use, to the point of becoming, in the XVI and XVII centuries, the obsession of the powerful. We mention, by way of example, Rudolf II, emperor of the Holy Roman Empire, who recruited the great astronomer Tycho Brahe for his horoscopes and, on his death, Johannes Kepler, considered a true specialist in divination; and General Albrecht von Wallenstein, a leading figure in the Thirty Years' War, who did not take a step without consulting his trusted astrologer, the Italian Giovanni Battista Seni, as Friedrich Schiller told us in his tragedy of the same name. Even Giovanni Cassini, first director of the Paris Observatory commanded by Louis XIV, had started his successful scientific career with a well-chosen horoscope. Then, the Enlightenment put a halt to this massacre of common sense, without however being able to stop the charlatans who continued to prosper, and still continue today, speculating on people's weaknesses and fears.

The *Commedia* contains frequent and precise astrological indications, used both in their most authentic meaning and as elegant and learned metaphors, linguistic expedients to decorate the poem. Perhaps the best known passage is found in canto XV of *Inferno*, where Brunetto Latini urged the poet to follow his birth constellation, that of Gemini, which was believed to favor intellectual attitudes, in order to grasp his literary and political objectives:

> *Se tu segui tua stella,*
> *non puoi fallire a glorïoso porto,*
> *se ben m'accorsi ne la vita bella;*
> * e s'io non fossi sì per tempo morto,*
> *veggendo il cielo a te così benigno,*
> *dato t'avrei a l'opera conforto.*
>   (*Inf.* XV, 55–60)

"If you follow your star, you cannot fail to reach a glorious port, if I perceived well during sweet life; and if I had not died so early, seeing the heavens so kind toward you I would have given you strength for the work."

Nothing surprising, given that in the didactic poem in the vernacular entitled *Il Tesoretto*, Dante's teacher had openly declared to believe that it was the will of "Almighty God" to give the seven planets and the Zodiac "*podere / in tutte creature, /secondo lor nature*" (canto X), "power to all creatures, according to their nature". A power that is a sort of obsession of the times and that Dante recognized, respected, and honored with his work aimed at looking with the eyes of the faithful to the work of God. *"Perhaps I will not see the light of the Sun and the stars?"*, he would later write in *Epistle XII* to an unknown Florentine friend who invited him to publicly repent in order to be able to return home from exile, *"Perhaps I will not be able to meditate on the sweetest truths everywhere under the sky?"*, to then add, with the usual haughtiness, *"Of course the bread will not be lacking"*, even if *"it tastes of salt*[2].*"*

In canto XXII of the *Paradiso*, Dante, who was born between May and June, explicitly refers to his zodiacal sign, Gemini: *"in quant'io vidi 'l segno / che segue il Tauro e fui dentro da esso"* (as I beheld the sign which follows Taurus, and was in it; vv. 110–111). This is not harmless calendar information, because a few verses later the poet invoked his *"glorious stars"*, those under which he felt *"the Tuscan air first"* and to which he owed his ingenuity, asking them to help him in accomplishing the last difficult part of his journey, the *"strong step"* that awaited him with the sight of God. As proof of the pregnant astrological value of these verses, Dante explained in the *Convivio* the importance of the heavens in the formation of the soul of men starting from conception:

> *E però dico che quando l'umano seme cade nel suo recettaculo, cioè nella matrice, esso porta seco la vertù dell'anima generativa e la vertù del cielo e la vertù delli elementi legati, cioè la complessione; [e] matura e dispone la materia alla vertù formativa, la quale diede l'anima [del] generante; e la vertù formativa prepara li organi alla vertù celestiale, che produce della potenza del seme l'anima in vita. La quale, incontanente produtta, riceve dalla vertù del motore del cielo lo intelletto possibile; lo quale potenzialmente in sé adduce tutte le forme universali, secondo che sono nel suo produttore, e tanto meno quanto più dilungato dalla prima Intelligenza è.*
> (*Convivio* IV, XXI, 4–5)

---

[2]This apparently strange sentence is part of the preference with which his ancestor Cacciaguida announced the exile:

> *Tu proverai sì come sa di sale*
> *lo pane altrui, e come è duro calle*
> *lo scendere e 'l salir per l'altrui scale.*
> (*Par.* XVII, 58–60)

"you will experience how salty tastes the bread of another, and what a hard path is to descend and mount by another's stairs." The bread tasting salt identifies lands and habits far from Florence's, where the bread is instead made without salt.

"Therefore I say that when the seed of man falls into its receptacle, namely the matrix, it carries with it the virtue of the generative soul, and the virtue of heaven, and the virtue of the combined elements, namely temperament. It matures and disposes the material to receive the formative virtue given by the soul of the generator, and the formative virtue prepares the organs to receive the celestial virtue, which brings the soul from the potentiality of the seed into life. As soon as it is produced, it receives from the virtue of the celestial mover the possible intellect, which draws into itself in power all of the universal forms as they are found in its maker, to an ever lesser degree the more it is removed from the primal Intelligence."

It would take a long time to list all the places in Alighieri's works that, explicitly or implicitly, allude to astrological references; so, let us show just a last couple of examples for everyone.

In the aforementioned initial triplet of canto XIX of *Purgatorio,* where the poet dreamed of the woman *"balba"* (stutterer), he mentioned the astrological properties of the two celestial bodies (Moon and Saturn) which were considered cold. In canto I of *Paradiso,* he spoke of the Sun rising at different points throughout the year. But at the spring equinox the *"lucerna del mondo"* (light of the world) is associated with the constellation of Aries — popping up where four circles (the celestial equator, the ecliptic, the equinoctial colure, and the horizon) intersect forming three crosses —, in a climatically and astrologically favorable condition for men:

> Surge ai mortali per diverse foci
> la lucerna del mondo; ma da quella
> che quattro cerchi giugne con tre croci,
>   con miglior corso e con migliore stella
> esce congiunta, e la mondana cera
> più a suo modo tempera e suggella.
>   (*Par.* I, 37–42)

"The lantern of the world rises to mortals through diverse outlets, but from the one that joins four circles with three crosses it comes forth with better course and joined to better stars, and it tempers and seals the waxy world more to its manner."

Also the Moon is used in a double symbolic/astronomical role. Virgil, in *Inferno* XX (vv. 127–129) addressed the poet saying:

> e già iernotte fu la luna tonda:
> ben ten de' ricordar, ché non ti nocque
> alcuna volta per la selva fonda.

"and already last night the Moon was full: you must remember it well, for several times it did not harm you in the deep forest."

In this passage Dante referred to an astronomical phenomenon, the full moon, and to a purely symbolic and astrological fact, the positive influence exerted on him by the queen of the night while in the dark forest.

Chapter 16

Conclusions

The family of one of us, Massimo Capaccioli, comes from a wonderful corner of Tuscany, the region of which Florence, Dante's beloved homeland remembered with vibrant words by Farinata degli Uberti in canto X of the *Inferno*, is the capital. It is called *Maremma* because it is close to the sea. Poor land, of clays, swamps, lagoons, scrubs, brambles, and bushes, once infested with malaria, where in a not-so-distant period the elderly, although generally uncultivated, recited passages from the *Commedia* by heart to the youngest gathered on the farmyard in the evenings sultry in summer. Sign of a root and an umbilical cord that should not be lost. Actually his father Anzio continued this tradition in his own way, as a great man of culture; he knew the whole *Commedia* by heart and always took the opportunity to quote and recite it.

Sperello di Serego Alighieri is a descendant of Dante, the nineteenth generation after the poet. Pietro Alighieri, son of Dante, lived in Verona, where Dante stayed for a long time, hosted by the Scaligeri. Pietro was also a judge in Verona and on April 23, 1353 he bought land in Gargagnago in Valpolicella, which is still owned by the Serego Alighieri family. However, the lineage from Dante is not entirely male, as in the XVI century there was only one female descendant, Ginevra Alighieri, who married Marcantonio di Serego in 1549. Francesco Alighieri, uncle of Ginevra and last male descendant of Dante, in 1558 bequeathed to Pieralvise, son of Ginevra and Marcantonio, various properties, including those in Gargagnago, provided that he and his descendants added to the surname of the Serego that of the Alighieri. The deed of purchase of the land in Gargagnago in 1353 and the will of Francesco Alighieri are kept in the family archives.

In a bookmark that Sperello received from his father Leonardo when he began to read, these lines of the *Comedy* are written:

> *O poca nostra nobiltà di sangue,*
> ...
> *Ben se' tu manto che tosto raccorce:*
> *sì che, se non s'appon di dì in die,*
> *lo tempo va dintorno con le force.*
> (*Par.* XVI, 1–9)

"O our paltry nobility of blood, [...] Truly you are a mantle that quickly shrinks, and, if we do not add from day to day, time goes around it with his shears."

In short, nobility and lineage vanish, prey of time, if we do not work to consolidate them. We hope with this volume that we have at least partially succeeded.

Index of cosmographic passages from the *Commedia*

We list out here the cosmographic passages that we quoted from the Commedia. The right column shows the page where they appear in this book.

Bibliography

Baldacci, O. (1965). I recenti contributi di studio della geografia dantesca, in *Cultura e scuola*, 14–15, pp. 213–225.

Baldacci, O. (1966). Alcuni problemi geografici di esegesi dantesca, in *Bollettino della Società Geografica Italiana*, 10–12, pp. 3–18.

Bettini, A. (2019). *Da Talete a Newton*, Torino (Bollati Boringhieri).

Billanovich, G. (1958). "Veterum vestigia vatum" nei carmi dei preumanisti padovani, in *Italia medioevale e umanistica*, I, Padova (Antenore), pp. 155–243.

Blair, M. (2015). *Cosmological Innovation in Dante's Divine Comedy*, Waco (Baylor University).

Boitani, P. (2017). *Dante e le stelle*, Roma (Castelvecchi).

Capaccioli, M. (2020). *L'incanto di Urania. Venticinque secoli di esplorazione del cielo*, Roma (Carocci).

Capaccioli, M. (2020b). Quando l'uomo scoprì le galassie, *Quaderni di storia della fisica* 1, pp. 1–41.

Capasso, I. (2017). *L'Astronomia nella Divina Commedia*, Pisa (Domus Galileiana).

Capocci, E. (2000). *Illustrazioni cosmografiche della Divina Commedia*, Napoli (Osservatorio Astronomico di Capodimonte, orig. edit. 1856).

Cordano, F. (2020). *La geografia degli antichi*, Bari (Laterza).

di Serego Alighieri, S. and Capaccioli, M. (2021). *Il sole, la luna e l'altre stelle — Viaggio al centro dell'universo dantesco*, Torino (GEDI).

Durling, R.M. (1996). *The Divine Comedy of Dante Alighieri*, New York (Oxford University Press).

Ferrari, A. and Pirovano, D. (2015). *Dante e le stelle*, Roma (Salerno Editrice).

Gizzi, C. (2017). *L'astronomia nel poema sacro*, Napoli (Loffredo Editore).

Le opere di Galileo Galilei, Edizione Nazionale, Firenze (Barbera).

Lucrezio Caro, T. (2003). *De rerum natura*, a cura di A. Schiesaro, trad. it. di R. Racanelli, Torino (Einaudi).

Oldroyd, D. (1986). *Storia della filosofia della scienza*, Milano (Mondadori).
Pecoraro, P. (1987). Nota cartografica, in *Le stelle di Dante*, Roma (Bulzoni).
Ricci, L. (2005). Dante's insight into galileian invariance, *Nature* 434, 717.
Virgilio Marone, P. (2005). *Georgiche*, in *Opere* a cura di C. Carena, Torino (UTET).

Made in the USA
Las Vegas, NV
02 September 2022

54547826R00105